THE MOUNTAIN FLOWERS
OF SOUTHERN TANZANIA

Grassland on the Kitulo (Elton) Plateau

THE MOUNTAIN FLOWERS OF SOUTHERN TANZANIA

A field guide to the common flowers

P.J.CRIBB
Royal Botanic Gardens, Kew

G.P.LEEDAL
Mbeya, Tanzania

Illustrated by
MAIR SWANN

A.A.BALKEMA / ROTTERDAM / 1982

In fond and grateful memory of
Mary Richards (1885-1977)

Prayer of Mary Richards

And you, mountains and hills, O bless the Lord.
And you, all plants of the earth, O bless the Lord.
And you, fountains and springs, O bless the Lord.
To him be highest glory and praise for ever.

from: The Canticle of Daniel

ISBN 90 6191 218 0

Printed in the Netherlands

CONTENTS

ACKNOWLEDGEMENTS

Many friends have helped us, both directly and indirectly, in the preparation of this book.

We would like to thank the following for the help and generosity during field work in Tanzania: Leonard Mwazumbi, our companion on many collecting grips, Dr Keto Mshigeni and Bob Wingfield of the University of Dar es Salaam; Lucy and John Tanner of Mazumbai, Rachel and Nigel Nicholson and Keith and Peggy Oliphant of Mbeya, Britt and Olle Alleljung, Maria and Pir Falk, Irene and Knut Johannson, Kirsten and Egil Hansen and Gerd Jorunn and Halvldan Jacobsen, all sometime of Uyole; Glory and Alan Holton and Maxine and Robin Humphries of Chivanjee; Mr and Mrs Kingdon of Mbozi; Mary and Jonathan Niblett, Liz and Alex Boswell, Debby and Mike Ghaui and Mr and Mrs Geoff Fox, all of Mufindi.

We are also grateful for the help, often considerable, of Miss Christine Kabuye and her staff of the East African Herbarium, Nairobi and to Professor J.P.M.Brenan, the Director, Mr Peter Green, the Keeper of the Herbarium and the following staff members of the Royal Botanic Gardens, Kew: Roger Polhill, Peter Taylor, Jeffrey Wood, Diana Bridson, Ray Harley, Martin Sands, Frances Davies, Susan Holmes, Dick Brummitt, Brian Mathew, Christopher Grey-Wilson, David Field, David Philcox and others. We would particularly like to thank Dr Bernard Verdcourt for so generously and meticulously editing the text and for kindly writing the introduction. Most of the colour photographs were provided by Christopher Grey-Wilson and that of *Disa hamatopetala* by Tony Gassner.

PREFACE

The foothills, plateaus and peaks of the Southern Highlands of Tanzania comprise one of the most interesting floristic areas in tropical Africa. Truly, here is a crossroads where the rich floras of eastern, southern and central Africa meet. In this mountain area, which stretches from east of Iringa through Mbeya to around Sumbawanga, upwards of 3 000 species of flowering plants can be found, probably over one-third of the total flora of East Africa.

Naming the plants of the region has always proved a difficult business. Access to *Flora of Tropical Africa* and to the published parts of the *Flora of Tropical East Africa* has not necessarily produced an identification. Probably the greatest success has been obtained by those with the patience required to send their pressed specimens to an herbarium such as those at Dar es Salaam, Nairobi or Kew. However, this is a laborious and time-consuming process. This book has therefore been written to fulfill the need for a compact volume which allows at least the showier flowers of the area to be readily identified. The arrangement of families followed is largely that outlined by Heywood (1978). We have not attempted to cover all the native species in this one volume – the sheer number and the fact that many are still poorly known or even, as yet, undescribed precludes this. Furthermore, the trees, larger shrubs, grasses and sedges have been excluded altogether because they are reasonably covered by other books as indicated in the bibliography.

It is less than a century since Goetze & Stolz, many of whose specimens proved to be of species new to science, first collected plants in the region. Recent collections have shown that much still remains to be discovered and large areas are still unvisited or have only been superficially explored botanically. Unfortunately, many of the most exciting areas are even now being destroyed by expanding cultivation or forestry. It would be a great shame if the character of areas such as the Kitulo Plateau changed before their rich floras were better known and safeguards enforced to ensure their survival for the enjoyment and enrichment of future generations.

We hope then that this book will be used as an introduction to this delightful and fascinating flora.

INTRODUCTION

The famous exploring expeditions with eminent scientists attached such as Banks & Solander, the Forsters, R.Brown, Darwin and Hooker passed East Africa by. It was not on the way to the South Seas, the spice islands, Eldorado nor even India and it was only the Cape which came in for extensive early investigation (see Gunn & Codd 1981). The East African coast had of course been visited by many Europeans at a quite early date. Even before Vasco da Gama sailed up the coast as far as Malindi in Kenya in 1497 other Portuguese are known to have been there a few years earlier. If they made any observations on the plants and animals in those early days they seem to have been lost. The first records seem to have been the few plants gathered by João de Loureiro (1715-1796) on his way back from Indochina in 1781-2. Other early collectors, mainly in Zanzibar, were J.Forbes (1798-1823), W.Bojer (1800-1856) and L.H. Boivin (1808-1852) but the first really important collection from the interior was made by Colonel J.A.Grant (1827-1892) during the famous journey when he accompanied J.H.Speke (1825-1864) during 1860-1863 on the second expedition to discover the sources of the R.Nile.

Since that time, literally hundreds of collectors have added to our knowledge of the East African flora but several of them have made outstanding contributions and added more than all the rest put together. Their names crop up all the time when one is working on a flora account, a revision or monograph, or merely doing routine naming. Among these notables are P.J.Greenway (1897-1980), B.D.Burtt (1902-1938), A.Peter (1853-1937), H.J.Schlieben (1902-1975) and, of particular interest for the area under consideration, Mary A.E.Richards (1885-1977) whose active botanical life spanned six reigns and only ended at the age of 92. When most folk are giving up or at least slowing down she decided to start afresh in Africa at the age of 65. With amazing industry she collected over 27 000 herbarium gatherings, mostly with duplicate material, and many Tanzanians will remember her energy and ability to outwalk and outclimb some over half her age. It is easy to account for her energy for she was a Victorian – that splendid era when everyone knew where they were going and what they wanted to do and got on with it unfalteringly.

Mary Stokes was born in Wales near Dolgellau but her parents' home was at Lichfield in Staffordshire. An early interest in nature was stimulated by her German governess but an expressed desire to seriously study botany at a university was unfavourably received by her parents, although she did attend some part-time classes. Mary married her cousin Major H.Richards in 1907 and they travelled the world extensively, she always taking an interest in the local flora. Her home was at Caerynwch near Dolgellau and here she brought up her family of one son and two daughters. After the 1914-1918 war, during which she turned her home into a Red Cross Hospital, she developed her interest in British field botany. Major Richards died in 1949 and next year Mary went to stay with the indomitable Gamwell sisters at Mbala in Zambia. Milne-Redhead, then at Kew, suggested she should collect, an idea she forcibly rejected, saying she was on holiday but, before long, a request for collecting equipment arrived and thus began her amazing new career in Africa for which she was awarded the MBE in 1969. From this base in Zambia Mary made many collecting trips into the Southern Highlands of Tanzania, mainly concentrating on Sumbawanga, but also visiting the Mbeya, Iringa and Njombe regions. She was particularly attracted to orchids and must have collected practically all the terrestrial orchids known in the Southern Highlands at one time or another. She finally left Africa in 1973 due to failing eyesight, having in the meantime moved to Tanzania but even though 88 she was collecting until the last. When she returned to Wales she amazingly set about the neglected garden of her old house Tynllidiart, of which she had regained possession, and returned it to something approaching its former glory, this time without any help from a gardener; and she could still walk five miles and tire younger companions.

This book, an introduction to the plants of the higher parts (above 1 500 m) of the Southern Highlands, is dedicated to Mary Richards, and Phil Leedal, one of its authors, writes 'Now she has gone back to the Elysian pastures to contemplate orchids and aloes undreamt-of on Earth. Her memory lives on inspiring us to do as she did, seeking the glory of the Creator in His flower world'. It is hoped it will enable many of the plants of the upland grassland and mountain rain forest, which she knew so well, to be identified from the illustrations and descriptions; needless to say with such a large flora one cannot deal with them all. For those whose interest has been aroused and who wish to go further a list of books is appended. One word of warning – although the collecting of Mrs Richards and others has been praised here and in fact our knowledge of the flora would be non-existent without collections, everything is now under so much pressure from increased agricultural and industrial development, concomitant with the increasing population, that plants should not be picked with abandon. There is no need for everyone to have a private collection; only a few are needed in schools, colleges and museums. The Southern Highlands are already a shadow of their former forested glory. Let us conserve what is left for all future Tanzanians and visitors to Africa.

Bernard Verdcourt
Former Director, East African Herbarium

1. GEOGRAPHY

1. *The mountains*

The natural feature known as the Southern Highlands since colonial times, is a great arc of high ground pivoting eastwards and westwards from the Mbeya region at its centre; eastwards and northeastwards towards and beyond Iringa and westwards and northwestwards to Sumbawanga. In the northeast, the limit is placed where the Luhombero massif falls precipitously at its northeast corner near Kidatu to the Kilombero River, while, in the northeast, the limit is formed by the low ground of the Karema-Katabe-Lake Rukwa trough.

Several distinct sectors can be distinguished in the Highlands (see figure 1). In the west, an uplifted rectangular mass of country around the town of Sumbawanga and between the rift valleys of Lakes Rukwa and Tanganyika forms the Ufipa plateau. It measures some 240 by 80 km in extent and lies mostly over 1 500 m altitude with higher plateaux at 1 850 and 2 100 m from which summits rise to 2 400 m or more. The principal summits working from the north are Masheta (1 910 m), Chala (2 350 m), Kate (1 986 m), Mtantwa (2 189 m) near Chala, Konkula (2 459 m) in the Mbisi forest near Sumbawanga, Malonje (2 416 m) southeast of Sumbawanga, Mmemya (2 228 m) (overlooking the Rukwa trough near Sakalilo) and Pito (2 346 m) and Kito (2 189 m), both near the towns of the same name. The dominant physical feature of the area is the great Rukwa escarpment which drops down from the Ufipa plateau into the trough where Lake Rukwa nestles at 792 m. To the west, the plateau falls in a series of steep steps to Lake Tanganyika (772 m).

This western sector of the Southern Highlands sweeps southeast without losing height into the central section centred on Mbeya, a large town with a population nowadays of over 100 000. Across the Tunduma or Msangano trough the land falls to about 1 250 m but then rises again to the Mbozi plateau, an area of high ground about 170 by 30 km in extent. This block of high ground extends from near Mkulwe southeast through Mbozi to the Bundali (Undali) Hills and then to the north end of Lake Nyasa. Its average height is 1 500 m and the highest points are Mbogo (2 456 m) in Umalila and Kisiba (2 309 m) in the Bundali Hills. The Umalila highlands swing round eastwards at high level to link up with the high ground of the Poroto Mts which average about 2 200 m with isolated volcanic peaks rising to nearly 3 000 m. The most spectacular of these are Rungwe (2 959 m) and Ngosi (2 620 m).

In their central section, one branch of the Southern Highlands swings south-eastwards following the shore of Lake Nyasa (475 m). This branch, the Livingstone Mts, is a spectacular range of peaks parallel to the Lake rising to over 2 400 m. One of the highest peaks of the range is Mulimuli which lies behind Lumbila on the lake shore, a town founded by the Germans as Allt Langenberg and their administrative capital in the area.

The other branch swings eastwards towards Njombe via the Kitulo (formerly Elton) Plateau and the bordering Kipengere Range. Here we have the highest ground of the Southern Highlands and the richest floristically as well. The plateau averages 2 400 m with the summits of Mtorwi (2 960 m) and Chaluhangi (2 923 m) appearing not much more than rolling hills around its margin. In this central section of the Southern Highlands there is also high ground north of Mbeya in the Mbeya range with Mbeya Peak itself rising to 2 824 m and scattered hills further north and northwest such as Isonge (1 707 m).

Between the Kipengere Range around Njombe and the eastern sector of the Southern Highlands the land falls to about 1 200 m in the region of Makambako. Then, to the north, rises the sloping plateau of Mufindi with its highest point at 2 070 m in the east before its drop sharply down the Mufindi escarpment to below 900 m. A few hills break the gentle expanse notably Irundi Hill (1 828 m). To the northeast of Mufindi the plateau merges into the Dabaga Highlands which rise to 2 140 m (Mt Ohuge). Westwards, the Mufindi and Dabaga Highlands drop slowly to the Great Ruaha River (c.700 m) with the town of Iringa sitting on a small hill at 1 602 m about halfway across the plateau.

2. *The rivers*

Several major rivers drain the Southern Highlands (figure 1). In the west the major river is the Kalambo which rises near Kate and flows southwards before pitching over the spectacular Kalambo Falls on its way to Lake Tanganyika. The Kafufu is the main river flowing into Lake Rukwa, which has no outlet. The Mtembwa drains southern Ufipa into the Momba River which also flows into Lake Rukwa.

In the central sector of the Highlands the large perennial rivers are the Kiwira, Mbaka and Lufirio, all draining southwards from the mountains into Lake Nyasa. The Nyasa Songwe drains the Mbozi block into Lake Nyasa while the Rukwa Songwe drains the Umalila and Mbeya Highlands into Lake Rukwa. In the Livingstone Mts (Ukinga) one of the largest rivers is the Lumbila which flows into Lake Nyasa, while the Ruhuhu and Ketewaka Rivers drain the Njombe Highlands southwards also into Lake Nyasa. The rivers flowing north from the central and eastern sectors of the Highlands all drain into the Great Ruaha River on the Usangu flats. Finally, the Little Ruaha drains the Dabaga Highlands and eventually joins the Great Ruaha.

3. *The lakes*

There are three major lakes in the area namely Lakes Tanganyika (772 m altitude), Rukwa (792 m altitude) and Nyasa (475 m altitude), the first and last being deep whilst Rukwa is relatively shallow. All have a fascinating fauna with large numbers of endemic fish and other aquatic life. A number of smaller lakes dot the Highlands such as Lakes Sundu and Kwela in the west. In the centre lies Lake Ngòsi (formerly Wentzel-Heckmann) filling a volcanic caldera while Lake Masoko near Tukuyu is another of volcanic origin. In the east, the Brooke Bond Leibig Tea Co. have dammed many small streams in Mufindi to create lakes, the largest being Lake Ngwazi of some 500 hectares.

4. *The roads*

The principal access to the Southern Highlands is via the all-weather tarmac road running from Dar es Salaam via Morogoro and Iringa to Mbeya and Tunduma and eventually into Zambia. It passes through the eastern and central sections of the Highlands and affords relatively easy access to a great section of the Highlands. An all-weather spur leads south from Uyole near Mbeya to Itungi, a port on Lake Nyasa. Elsewhere all the roads are earth- or grit-surfaced and are subject to the vagaries of the weather and the heavy traffic using them. Particular care is needed on all of these roads especially in the rainy season.

2. GEOLOGY

It is said that no botanist intent on studying the flora of an area should be without some knowledge, even though slight, of the geology of his area. Ultimately, it is the geology that has controlled the distribution of much of the plant life of the Southern Highlands since it is the geology that has influenced so profoundly the soils and climate of an area. The Rift valleys formed the lakes and the lakes (notably Tanganyika and Nyasa) form their own climate; thus the head of Lake Nyasa is the wettest area in East Africa with 285 cm of rain per year and the correspondingly rich and lush vegetation. In contrast, in the rain shadows of the Southern Highlands, on the plain areas of Ruaha, Usangu and Rukwa there is a very dry climate with a corresponding savannah vegetation.

Geologists distinguish three great groups of rocks namely Igneous, Sedimentary and Metamorphic. All three groups of rocks make up the Southern Highlands. The first rocks to form in this area when the earth's crust cooled were granitic rocks later partially altered by heat and pressure movements when mountain building took place some 3 250 m.y. ago. These ancient granites and gneisses occupy the whole heartland of Tanzania. The Southern Highlands have been wrapped round this shield area by succeeding depositions of sediments on the margin of the shield (the sediment being derived from mountains of the shield areas) and their subsequent metamorphosis. A sedimentary rock is simply one that has been formed by deposition of material in water (lakes, oceans or geosynclines as geologists call them) and then the sediment is subsequently hardened by compaction due to the weight of overlying sediment. Such sedimentary basins formed on the margins of the Dodoman central shield and sediments of the Ubendian (U) were deposited in them. These geosynclines were ultimately folded by movements of the earth's crust (plates) and the sediments in them, many thousands of metres thick, were folded and metamorphosed by heat and pressure and raised up into high mountains (the Ubendian Mountains orientated NW-SE). The new mountain range was slowly over millions of years eroded away. Geologists have dated this formation of the Ubendian mountain range as about 1 800 million years ago. Their eroded remnants now form much of the high ground in the West and Central area of the Southern Highlands. The botanist will encounter typical rocks of this period in Ufipa, Undali, the Mbeya Range and Ukinga. They include such typical rocks as gneiss (a type of foliated granite), schists, and marbles, are always crystalline and have what is called layered or foliated appearance. Widespread granites were intruded at this time.

Lying unconformably on the top of these ancient crystalline metamorphic rocks are rocks called the Bukoban (B). 'Unconformably' means, as its name implies, that there is an age gap between the Bukoban and Ubendian. Indeed the great series of sandy Bukoban rocks were laid down for the most part in desert conditions about 550 million years ago. The old mountains were being worn down in dry conditions and their rock material was deposited in dry or occasionally shallow-water basins. These sandstones are particularly conspicuous in the central Southern Highlands in Buanji where miles and miles of country looks like a slag heap with finely divided slate material covering the ground and no grass growing on it even in the wet season. These sandstones are also responsible for many waterfalls in the Southern Highlands, notably Kalambo and Salala (Igurusi-Chimala).

After this the Highlands area remained reasonably quiet for some 400 million years and no new rocks were formed. The old mountains were all worn down and no seas transgressed over the area. But about the beginning of the Permian age, some 270 million years ago, lake basins began to form in troughs across our area; these basins probably were initiated in Pre-Cambrian times when the Ubendian was folded for they follow the general NW-SE trend of the Ubendian folding. Sedimentary rocks, sandstones, shales and even coals were laid down in the troughs during the Permian to Cretaceous period (270-70 million years ago). Remnants of these sediments occur in the borders of the rift valleys in the Rukwa and Unyakyusa. Important for the botanist during this period is the Panda Hill limestone or carbonatite complex, a volcano of some 110 million years ago.

Faulting, tilting, uplifting and downwarping and sinking of the Southern Highlands continued on a small scale during this period (270-70 m.y.). In the period from the end of the Cretaceous important movements generally known as the Rift Faulting of the earth's crust occurred in the area leading to the formation of the Rift Valleys and the large lakes, the outpouring of great masses of lavas and the building of volcanoes. In the Southern Highlands, two great rift systems are distinguished: the one in the west in which Lakes Tanganyika, Rukwa and Nyasa lie is called the Western Rift; the one in the east the Eastern Rift. The lavas and volcanoes are found mainly in the Rungwe area and extend almost down to Lake Nyasa. Ash deposits occur in soils from Mbozi eastwards to Ubena.

The final rocks of our area are the Lake Beds, a group of sedimentary rocks (sandstones, silts and laminated clays) occupying former lake basins and dating from about the Miocene to Pliocene age (25-2 million years ago). The Usangu trough, the Rukwa and Great Ruaha basins and the Kyela plain are filled with these sediments.

3. CLIMATE

If you drive up from Dar es Salaam to Mbeya at the end of April you will no doubt meet some rain on the way and the air will certainly be warm and sticky. The country will be under a cover of luxuriant vegetation with the trees in leaf, the grass green and an abundance of flowers everywhere. This is the height of the wet season which in the Southern Highlands is also the warm season. It is, in fact, Tanzania's summer with the sun directly overhead at mid-day from March until May. A warm wet belt of air follows the sun and lies over the Southern Highlands between about late November and late May causing the so-called wet season ('masika' in Swahili). The winds during this period are variable mainly easterly or northeasterly but occasionally westerly. As this belt of warm wet air moves south into Zambia in January there is a slight lull in the rains until March when, with the return of the sun overhead, the rains recommence in earnest.

During the wet season the rainfall is variable throughout the region. At the head of Lake Nyasa, which is probably the wettest place in the whole of the country, an average of 285 cm falls each year. In some places in the Poroto and Livingstone Mts as much as 400 cm has been recorded. In four consecutive days in 1972 0.5, 20.1, 20.1 and 7 cm were recorded there and on 9 April 1955 42.5 cm of rain fell on the slopes of Mt Rungwe causing widespread landslides. This explains why much of the higher and some of the lower areas of Unyakyusa are covered with evergreen or bamboo forest.

Other areas of the Southern Highlands receive much less rain. Lugoda, for example, which is the wettest part of Mufindi averages 114.8 cm a year. Moving westwards from this station, the rainfall falls off progressively until at Madibira (1 159 m) it averages only 60.3 cm a year. Here in the plains the dry season lasts for six months and this is true over much of the lower ground to the north of the Southern Highlands. The rainfall pattern in Ufipa is best illustrated by the two towns of Sumbawanga and Kala. The former at 1 723 m altitude averages 82.3 cm falling in 78 days while the latter, on Lake Tanganyika, averages 119.9 cm in 67 days. However the year to year variation is great with Sumbawanga's rainfall dropping as low as 55.5 cm as compared with a high of 114.9 cm. Similarly, at Kala, rainfall has varied from 62.5 to 308.6 cm in different years.

The journey from Dar es Salaam to Mbeya in October is quite a different experience. On climbing into the Southern Highlands near Iringa you will immediately notice that the land is parched, the trees now leafless, the grass withered and the rivers mostly dry or very low. This is the dry season ('kiangazi' in Swahili) and is not the best season for botanists. The warm, wet air-belt is now away north over Kenya and the southeast monsoon winds are blowing across Tanzania from June until October. These winds are dry and dessicating and quickly dry out the land. It is also the cool season. On the Kitulo plateau at 2 850 m a temperature of –5°C has been recorded and night frosts are not uncommon. Once every 25 years or so, night frosts occur all over the Southern Highlands with devastating consequences to the vegetation.

4. VEGETATION

The traveller in the Southern Highlands will recognize that there are a few broad well-defined types of vegetation such as the hot almost desert-like lowlands (at least in the dry season) of Usangu and the rather wet cold and cloudy mountainous areas covered by dense rain forest such as Mount Rungwe, north of Tukuyu. Another vegetation type he should be able to distinguish is the monotonous woodland called 'miombo' in Swahili such as in the country bordering the Dar es Salaam to Mbeya road as it approaches Iringa from Morogoro. Broadly speaking, the vegetation zones follow the height of the land, with savanna and semi-desert vegetation zones in the low-lying areas grading at about 1 300 m altitude into the woodlands and these in turn giving place to the upland grasslands and forest patches on the higher ground. If you stop almost anywhere in the Southern Highlands and look around you should be able to pick out the woodlands at the intermediate levels and then on the higher ground the open grasslands interspersed with forest; thus, at Iringa, one is in the woodland belt (now largely cut down) but if you look eastwards you will see on the higher ground open grassland areas and patches of forest and the same is true at Mbeya and Sumbawanga.

We can, for convenience, distinguish 12 major types of vegetation in the Southern Highlands which should slowly become obvious to observers in the area. Working from the lowest areas we start of course with the great lakes, Nyasa, Tanganyika and Rukwa clearly shown on the map (figure 1). Lakes Nyasa and Tanganyika have a regular shape being very deep and their margins are fairly constant (there was a remarkable rise of 2 m in the lake level in April 1979 which caused a lot of semi-permanent flooding). Lake Rukwa however is a shallow lake and its margin is constantly changing. Its present borders are the highest ever known due no doubt to continued heavy rainy seasons since the low of 1950-1955. The floating islands (sudd) of the water lettuce (*Pistia stratiotes*) are the only visible vegetation on the lakes particularly in Lake Rukwa. They form at the margins of the lakes and later break free during high winds.

Around the shores of the great lakes is our second vegetation zone namely permanent swamp with water present throughout the year. The vegetation here consists of grasses (in the Rukwa area the most frequent being *Vossia cuspidata*), reeds, rushes and occasional ferns, the most conspicuous element being papyrus (*Cyperus papyrus*); it also occurs in marginal swamps of lakes at higher levels for example around Lake Ngwazi at Mufindi.

Outside the permanent swamp areas surrounding the lakes are extensive areas of flood-plain grassland, flat, low-lying areas which fill up with water (overflow from lakes or rivers) in about December or January (the early rains) and continue to be swampy well after the rains end in early May into June and even July. These flood-plain grasslands occur extensively round Lake Rukwa and at the northern end of Lake Nyasa. The dominant grass in the North Rukwa plain is *Echinochloa pyramidalis.* Palm trees emerge from these grasslands, the deleb or Palmyra palm, *Borassus aethiopum* ('mvumo' in Swahili) and the doum palm, *Hyphaene ventricosa* ('mkoche' in Swahili). *Acacia* trees are also a common element here.

Usually lying above the flooded areas of grassland are areas of wooded grassland which may occasionally be flooded but, on the whole, even in the wet season remain unflooded. This is the typical so-called savanna or parkland widespread at about 700-1 000 m in the Usangu plains and the Rukwa trough, technically defined as having trees less than 50 % of the ground-cover although, in fact, great expanses are often treeless. Any of the roads leading off west or north from the main Iringa to Mbeya road will eventually take one down into this dry savanna country but the usual route into Usangu is via either Rujewa or Igurusi-Utengule. A bewildering variety of trees occur here with the genera *Acacia, Sterculia, Commiphora, Grewia, Capparis* and many others being well represented. The robust perennial grasses *Panicum maximum* and *Digitaria* are also prominent features of this vegetation type.

The savanna country gives place to woodland at about 1 000-1 300 m in the Rukwa trough, Usangu and elsewhere. This woodland is the famous 'miombo' country of East and south-central Africa. It covers an enormous area of country in the Southern Highlands, in fact most of the land lying between about 1 200 and 1 800 m, including a large part of Uhehe, Upangwa, Unyakyusa, Chunya, Rukwa, Ufipa, Mbozi and the area around Lake Tanganyika. In these areas the tree cover is well over 50 % and the dominant trees are those belonging to the genera *Brachystegia* and *Julbernardia* and to a lesser extent *Isoberlinia.* With a bit of practice one soon sets used to recognizing the *Brachystegia* trees. 'Miombo' is simply a common Swahili and tribal name for them. These are at their best in October and November when the new leaves are bursting out in brilliant flushes of red, bronze, yellow and green. In the dry season the trees are for the most part leafless in these deciduous woodlands.

Bushland and thicket, a sixth vegetation type, is defined as land carrying more than 50 % cover of densely growing evergreen or deciduous shrubs or small trees. Some of these thickets may be impenetrable with one's way blocked by dense masses of thorn trees. The whistling thorn (*Acacia seyal*), for example, can form dense almost impenetrable thickets on the Usangu plains. Thicket country can develop either in the woodlands or in the savanna.

Throughout the Southern Highlands, the 'miombo' woodlands are replaced by upland or montane grasslands at about 1 800 m. Sometimes this boundary is well-defined but usually woodland trees continue to struggle up the mountainsides forming a gradation. The grasslands are perhaps the most beautiful places in the Southern Highlands. The vegetation of these areas is predominantly fire-climax grassland of medium height. Towards the end of the dry season, in October and November, vast areas of these grasslands are burnt. It is lovely country, easy to walk over and often with wonderful vistas. Dominant genera of grasses are *Andropogon, Eragrostis, Hyparrhenia, Pennisetum* and *Setaria.* Additionally, they are always covered with flowers even in the dry season.

The upland grasslands are found right up to the summits of the mountains at nearly 3 000 m. Scattered across them, particularly in wetter situations and valleys, are forests and patches of forest. This is the montane rain forest easily recognized from a distance by its dark colour against the grasslands. To enter the forests from the grasslands is like entering a new world. Giant trees such as *Aningeria* and *Chrysophyllum* can tower to 50 m. Lianes are everywhere and smaller trees form a lower level canopy whilst the floor of the forest can be densely covered with shrubs. A great number of tree species may be present in any one forest. In wetter situations, often on slopes, the rain forest may be interspersed with bamboo (*Arundinaria alpina*) or the bamboo may take over completely to form its own forest. The floor of the forest is often rich in shrubs and herbaceous plants particularly near the margins. Epiphytic orchids are also well represented and can be found growing on tree trunks, branches and twigs often with their roots covered in mosses and lichens.

Throughout the woodlands are upland grassland areas where the ground is flat and liable to seasonal flooding and swamp or marshy areas are found. These are species-rich habitats and in the rainy season are carpeted with flowers being particularly rich in terrestrial orchids. These swamps are called 'mbuga' in Swahili. They are particularly well-developed along the road between Iringa and Makambako where several large ones can be seen. Another large series occurs in Mbozi on the roads from Mlowo to Mkulwe and Mlowo to Vwawa. At higher altitudes, smaller swamps can be found on the Kitulo Plateau and in Ubena. Many of these never dry out and are worth visiting throughout the year.

The rock deserts, a relatively small but important vegetation zone, are found on the recent volcanic mountain summits, such as Rungwe, where there is little vegetative cover apart from mosses, lichens and grasses. The ponds, lakes, streams and rivers of the area, for example, the dams in Mufindi, are well worth searching for plants. The larger rivers are bordered by a narrow forest belt, riverine or gallery forest, which as the altitude decreases may penetrate well into the woodland zone. Orchids, particularly epiphytic ones (*Aerangis, Angraecum* etc.), are frequent in such places often growing on trees overhanging the water.

Finally, we must consider the large areas altered by man. Cultivation in some areas is extensive and rapidly increasing in extent. Furthermore, woodland and forest are diminishing year by year as the trees are felled for their timber, for charcoal or for cultivation. The plants found in such areas are frequently weeds associated with cultivation and are often widespread species differing from the native ones. When a previously cultivated area is deserted then a secondary vegetation arises which again often bears little resemblance to the original flora.

These then are the main vegetation zones of the Southern Highlands. Slowly any traveller in the area will learn to recognize them and thus appreciate all the more the beauty of the Southern Highlands. As we have stated the majority of the flowers described and figured in this book are mountain flowers, that is, they are found in the areas above the woodlands in the upland grasslands and forests usually above 1 800 m.

5. TRIBES AND PLANT NAMES

The Southern Highlands for administrative purposes is divided into three regions ('*mkoa*' in Swahili), Iringa, Mbeya and Rukwa. Each region is further subdivided into districts ('*wilaya*' in Swahili). Iringa Region has the districts of Iringa (rural and town), Mufindi, Njombe and Rudewa. Mbeya Region comprises the districts of Mbeya (urban and rural), Mbozi, Rungwe, Kyela, Ileje, and Chunya. Rukwa (Sumbawanga) Region contains the districts of Sumbawanga, Mpanda and Sumbawanga town. The two regions of Mbeya and Iringa have more or less the same population of about a million, with Rukwa Region more thinly populated at 300 000. Densities of population range from about 10 per km^2 in Rukwa region to 15 and over in the Mbeya and Iringa Regions. Normally the higher or wetter areas are more densely peopled, for example, Unyakyusa where the density averages 100 people per km^2. The increase in population is tremendous with an average of about 3 % per annum in the rural areas and 10 % or more in the urban areas. The growth of the towns is rapid and likely to continue, most of the major ones such as Sumbawanga, Mbeya and Iringa having leapt from a population of a few thousand before independence to 50 000-100 000 now.

1. *Iringa region*

The Iringa Region is peopled by the Hehe, Bena and Pangwa tribes. The Wahehe (in Swahili the prefix Wa- refers to the tribe and the prefix Ki- refers to their language, whilst the prefix U- refers to their country of domicile; thus the Wahehe speak Ki-hehe and live in U-hehe country which covers the districts of Iringa (rural and urban) and Mufindi). Smaller groups of other tribes are of course to be found everywhere; thus Wakinga and Wabena will certainly be found in Uhehe where they have gone for work. W.Carmichael, who was a forestry officer in the late colonial times (1949-54), collected a large number of Hehe names for plants. Like almost all the tribes in the Southern Highlands the Wahehe have names for most of the trees and for many of the shrubs and other plants which are utilized. Decorative plants which have no use are not named. Like all tribal names for plants some of the Hehe names are very musical and of real value for identifying species; for example 'munongo' the name for *Sclerocarya caffra* is similar to that used by other tribes for this tree. Many other collectors have amassed Kihehe names for plants. South of the Wahehe are the Wabena who are grouped with the Sangu, Hehe, Pangwa, Kinga, Wanji and Kisi people in the same language group. The Wabena are a numerous tribe right across the Njombe Highlands giving place further to the south and south-west in the Rudewa district to the Wapangwa. Along the shore of Lake Nyasa are the Wakisi. In the high mountains of the Kipengere to Bulongwa are the Wakinga. Dropping to the plains of Usangu along the Great Ruaha River one encounters the Wasangu. These people have a great number of names for the savanna trees and shrubs of their area and a large number of these have been collected notably by P.Greenway. Much of the Usangu area lies within Mbeya Region.

2. *Mbeya region*

The people of the Mbeya Region belong to a slightly different language group from those of Iringa-Njombe. They are grouped linguistically into two zones, one embracing the Southern Highlands tribes of Wanyakyusa, Wanda, Wanyam-wanga, Wanyiha, Wamalila and Wasafwa and the other including the Rukwa area tribes of Wakimbu and Wabungu.

The Wanyakyusa tribe of nearly half a million people is one of the largest and most influential tribes of Tanzania. They occupy the low ground in the watershed of upper Lake Nyasa. We know quite a lot of Nyakyusa plant names thanks to the efforts of various workers. The flora here is very rich and in many ways remarkable owing to the high rainfall. Lying immediately along the high ground surrounding Unyakyusa are the mountain tribes of the Wandali, Walambya, Wasafwa, Wamalila and Wanyiha. The Chunya District is inhabited by Wabungu and Wakimbu people, the Mbeya district by Wasafwa and Wasangu; the Mbozi district by Wanyiha and the Ileje District by Walambya and Wandali. Kibungu and Kikimbu have much in common and the plant names are often identical.

3. *Rukwa region*

The Wafipa, comprising well over half the total population of the region, are the dominant tribe here, but amongst them are small pockets of Wanyiha. South of the Wafipa are the Warungu and Wamambwe, and north and northwest are small tribes like the Wapimbwe and Warungwa in the Rukwa trough.

PLANT NAMES AND USES

After describing the plants, we have given, where known, the local name or names. Several hundred tribal names of plants have now been collected by various workers and, as we have said, the vast majority of these names apply to trees and useful plants. It is rarely that one can get local names for plants with no local use. However, if one can find a local medicine-man who has a deep interest in plants for medicinal or magical reasons he will usually be able to name practically anything. Just as the languages resemble one another, so too the names of plants are often similar from tribe to tribe; thus *Catha edulis* (a small tree) is 'nzuruti' in Malila, 'nsuluti' in Ndali and 'insuluti' in Nyakyusa. The 's' and 'z' and the 'l' and 'r' are often interchangeable from one language to another. Trees will practically always have a name in the local language but one may need time and patience to get it. Very often one can only carry a branch of the tree for naming rather than show the tree to the namer and this places the local man at a disadvantage. It is always best to try and consult a group of people (old men are the best) rather than one person for many obvious reasons. With regard to the medicinal uses of plants we have mentioned little in this account. Practically everything has been tried or used for medicine and there can be little doubt of the effectiveness of some plants. A useful summary of East African plants of medicinal value is given by Kokwaro (1976).

6. HIGHWAYS AND BYWAYS OF THE SOUTHERN HIGHLANDS

Most people travelling through Tanzania have with them the Shell Map of East Africa (1965, George Phillip and Sons Ltd, London) on the scale of 1:2 000 000. As will be seen from this map, the main access road is the 'Tanzam Highway' built in the early 70's to link Dar es Salaam with Zambia (hence the name). The Shell map shows the pre-Tanzam alignment of the main road which, however, was fairly closely followed by the subsequent road builders. The distances shown on this map more or less agree with the present-day distances along the new alignment. For our purposes we count the Southern Highlands as beginning at Iringa (Dabaga Highlands). From Iringa the Tanzam heads more or less in a straight line southwards to Makambako before turning westwards to Mbeya and on to Tunduma at the Zambian border. From Tunduma, a secondary road (unsurfaced) leads off to Sumbawanga in Ufipa and onwards to Namanyere where the western branch of the Southern Highlands finishes. The Iringa to Tunduma road is now surfaced and all-weather, although care has to be taken on account of bad patches in places and enormous Juggernaut lorries on their way to Zambia on what was once called the 'Hell-run' and still is in many ways! Be on the look-out also for cyclists who for some reason or other best known to themselves ('better to see what bumps into you from the front than not see what hits you from the back') travel on the wrong side of the road (this especially from Makambako onwards). Numerous branch roads of varying quality lead off this main road. The Sumbawanga road in the dry season (May-November) is passable through roughish; in the wet season (December-May) it is decidedly a road for a four-wheel drive vehicle as it can be sticky for miles. Hopefully it will be surfaced soon as all future development of the new region of Rukwa is dependent on this road being improved. Again, there are numerous secondary roads leading off this central one. Outside the tarmacked Iringa-Tunduma stretch the only other surfaced road is that built by the Holzmann company in the late 60's linking Uyole (east of Mbeya) with Itungi Port on Lake Nyasa (referred to as Lake Nyasa in Tanzania; on the Shell map it is called Lake Malawi).

For the purposes of describing access to the Southern Highlands and places of special botanical and other interest to the general or specialized traveller we cannot do better than follow the late Mrs Mary Richards on her own travels through the Southern Highlands; she managed to get into most corners of these mountain areas and some of her travels have been written up by W.Condry in his *Birds and Wild Africa.* In following Mary in this way, we have extracted information on her travels from her botanical notebooks.

Mary Richards made her home at Abercorn (now Mbala) in Zambia. From 1951, she collected plants in that area summarizing these in her cyclostyled book *Check-list of the flora of Mbala (Abercorn) and District* which must surely include most of what is to be seen in that area. From Abercorn, she went off on a series of collecting trips in a Land Rover driven by the late Ali Omari and, later, Modesty Sanane. The plants she collected are kept at the Herbarium of the Red Locust Control Service in Mbala and duplicates of these are at Kew.

1. *The Ufipa Highlands*

On her first recorded foray into Tanzania in late November 1954 she visited the Mbizi (or Mbisi) forest. The road from Mbala (Abercorn) crosses the border at Kasesya (24 km) then passes through Isopa on its way to Sumbawanga (112 km). From Sumbawanga, a rather dangerous road climbs into the hills up to the Mbizi Forest, rising to 2 454 m at Konkula and one enters the forest 12 km from Sumbawanga. It is noted as a locality of the rare Red Colobus monkey. Flowers seen in the forest and on its margins in November include numerous composite species (*Eupatorium, Vernonia, Nidorella, Gerbera, Sonchus* and other genera), *Hibiscus rhodanthus, Coleus scaposus* (Labiatae), *Hypoxis goetzei* (Hypoxidaceae) and *Scilla indica* (Liliaceae). It is at this time that the rains are breaking in the Southern Highlands and it is the best time for seeing the early flowers.

From Mbizi, Mrs Richards went to Malonje, which became one of her favourite localities, where there used to be a farm owned by Mrs v.Dam. Malonje Mountain (2 411 m) is 16 km southeast of Sumbawanga. The open grassy slopes at this time of the year (November) are alive with flowers such as *Scutellaria pauciflora* (Scrophulariaceae), *Cyanotis speciosa* (Commelinaceae), *Pentanisia arenaria* (Rubiaceae) and *Vernonia natalensis.*

After Mbizi and Malonje one travels to Mpui (64 km from Sumbawanga) and Lake Kwera (16 km further to the east). This beautiful little lake was a favourite spot for Mrs Richards and always makes a welcome retreat for travellers from Tunduma to Sumbawanga.

From Lake Kwera it is a short trip to the Sakalilo escarpment on the east where the land falls precipitously from Mmemya (2 230 m) to the Rukwa plains at 822 m. Typical plants to be seen in the Mmemya are *Begonia princeae,* the lovely tree *Stereospermum kunthianum,* the asclepiads *Gomphocarpus amabile* and *Ectadiopsis oblongifolia,* and *Droogmansia whytei* (Leguminosae). The views from the top on a clear day are spectacular.

In September of 1956 Mrs Richards collected in the Kito area of Ufipa. Kito and Kito Mountain (2 189 m) lie 16 km north of the border on the road from Mbala to Sumbawanga. September in Ufipa is rather dry and much burnt but many flowers can still be seen such as *Aster tanzanicus, Wahlenbergia virgata* (ubiquitous in the Southern Highlands), *Helichrysum latifolium* (an everlasting flower), *Pentanisia sykesii, P.schweinfurthii* (a widespread blue-flowering Rubiaceae flowering early in burnt ground), *Moraea thomsonii* (an iris) and the pretty *Monadenium orobanchoides* (Euphorbiaceae).

In the same year but in November she returned to Lake Kwera collecting there the lovely water lily, *Nymphaea caerulea.* Mary described its flowers as pale blue on stalks held several cm above the water. Another aquatic collected by her there was *Utricularia vulgaris.* On Mmemya mountain to the east of Lake Kwera she found the orchids *Nervilia kotschyi* and *Eulophia thomsonii.*

Some 25 km north of Sumbawanga is the Muse (or Muze) area and here Mrs Richards collected *Kaempferia rosea* towards the end of November. Further along the road in the direction of Mpanda is Chala and Chala Mountain (2 349 m) 64 km from Sumbawanga and here, in early December, grow such flowers as *Kniphofia, Pentanisia prunelloides* and *Begonia tayloiana.* On 13 December (1956) Mary collected the outstanding *Gladiolus melleri* near Sopa village, 12 km NE of Chala. Then, as the New Year approached, she moved on to Kasanga which is 40 km from Isopa and lies on Lake Tanganyika, crossing the Kawa River. At the Kawa Falls, she made extensive collections including several rare Asclepiads.

In March 1957 she returned to Malonje collecting orchids such as *Habenaria uhehensis* and *Pteroglossapsis carsonii* and Rubiaceae such as *Spermacoce dibrachiata* and *Galium scabrellum.* A trip was then made to the Kalambo River and Chapota (56 km from Sumbawanga). Roads here are bad and best avoided in the wet season without four-sheel drive transport. Many Leguminosae were collected such as *Aeschynomene schimperi, A.nyikensis, Indigofera homblei* and *Crotalaria dasyclada.* In the Chapota swamp, she collected *Bidens richardsiae,* a new species named after her. On the 8th she collected in the Ipeta swamp near Chapota; on the 9th she was at Kasapa, just west of Chapota; then on the 12th back to Malonje; on the 13th in the Mbizi Forest again and later she returned to collect extensively around Malonje, Lake Kwera and the Ilemba Gap.

After the wet season, Mrs Richards ventured down the road to Kasanga (13th to 21st June 1957). Among her many collections here were legumes such as *Cassia nigricans, Tephrosia linearis, Vigna luteola* and *Crotalaria senegalensis.* This lovely place is the site of an old German boma but it is, however, full of mosquitoes in the wet season.

In November of 1958 she returned again to Kito Mountain collecting *Dierama pendulum,* the orchid *Nervilia kotschyi,* *Sansevieria kirkii* on anthills, and the lovely little *Scilla rigidifolia,* conspicuous for its blue stamens against the white petals.

In December, her party arrived at Lake Sundu lying south of the road from Munza to Mwazye. Collections here included the trees *Acacia polycantha* and *A. sieberiana* and the shrubs *Acalypha ornata* and *Carissa edulis* with its white, sweet-scented flowers. Mchata Mountain (2 284 m) to the north was also visited and further botanising around the Kalambo River ended the collecting for the year.

In February 1959, Mary was again in the Kawa River area and Kasanga and then later went on to the Kanda Hills near Sumbawanga and Lake Kwera again. On the Lake, she collected *Potamogeton schweinfurthii* and near the edge *Hygrophila auriculata,* then on to the Ilemba gap where she collected *Trichodesma zeylanicum.* In March, the party descended via the Muse gap to the Rukwa plains (c.800 m) passing Namwele; this is a difficult road.

She made two further trips the following year to the area in July and August visiting the Nsangu forest-Mole area and the Mbizi Forest and, in November, Kito Mt and Lake Sundu. In April 1961, she returned to Kito and later the same year visited the Namwele escarpment, descending it to the Nziga plains on the Rukwa. On the last day of the year, she revisited the Nsangu Forest where she saw the mauve pea *Vigna monophylla.* By early January the party had descended to the Rukwa again collecting the water fern *Azolla nilotica* on the Nziga plains. Then on to the Bonta plain at the mouth of the Kavu River. This is an unspoit area of wild Africa well described in Condry's book. On 12 February she reached Lake Katavi on the north side of the famous Liande sand ridge. The roads here are very bad and of course only passable with Land Rover and similar vehicles. Again many interesting aquatic and flood-plain plants were collected, most of them growing in watery sands, such as *Dioscorea schimperana, Tephrosia lupinifolia* and *Ricciocarpus natans.*

In December 1963 she visited Kala Bay and Kirambo on Lake Tanganyika. The mission at Kala is a superb spot offering wonderful views over the Lake from which alone it is accessible. Here at the Lake edge she collected in 'miombo' woodland *Polygala sphenoptera* and many woodland trees and shrubs such as *Commiphora caerulea* and *Clerodendron myricoides.* A visit was also paid to Malasa Island just off the coast.

Kalambo Falls, one of the great sights of Tanzania where the Kalambo River drops over a vertical escarpment of sandstones of the Buanji Series was visited several times during Mrs Richards sojourn at Abercorn (see W.Condry's book). They are only accessible from Mbala and above them is a Stone Age site.

Malonje was visited for the last time by Mrs Richards in October 1965. The farm was later closed and she herself left Zambia to live at Momela Lodge near Arusha in northern Tanzania. In Ufipa, thanks to its proximity to Mbala, she collected thousands of flowers during the safaris outlined above. Anybody following in her footsteps will certainly find some of the plants she saw. The main areas of interest in Ufipa and Rukwa have been outlined above, places like Kalambo Falls, the Mbizi and Nsangu forests, the Malonje Plateau, Lakes Sundu and Mwera, Kito, Chala, Mmenya and Sakalilo Mountains, Lake Katavi, the Liande Ridge, the Sopa and Nziga plains and indeed much of the area is still unspoilt and beautiful flower country.

2. *The Mbeya Highlands*

The Mbeya Highlands (fig.2) include the country southeast from Ufipa extending eastwards to the Njombe highlands. Mrs Richards made several safaris through this area but never collected as extensively as in Ufipa which was on her doorstep so-to-speak!

These highlands undoubtedly form some of the most beautiful scenery in the world, due to the diversity of their geological history, climate and altitude. Access is from Mbeya which does indeed lie in the heart of these mountains at an altitude of 1 700 m, a superbly situated town lying below Mbeya Mountain (2 824 m). To the south is the great volcano of Ngosi (2 620 m) and on the east the great bulk of sandstone called Ishinga Mountain (2 687 m).

Since the time of Mary Richards the Iringa to Tunduma road has been tarmacked as has also the Mbeya to Itungi Port road. These two highways greatly help visitors to the area. Some of the other roads she traversed are still just as they were! Of special note is the amazing road with 55 bends which climbs up the Chimala escarpment which even has a bus along it!

Mary first visited these Highlands on 14 October 1956 following in the steps of a large number of collectors. She collected sporadically at Tunduma, Mbozi and the Mbeya Range before going on to Rungwe.

Mbozi is an interesting area beginning more or less at Tunduma and extending across the so-called Mbozi plateau (or Horst block to the geologist) northwards to the escarpment that falls down into the Songwe valley and Songwe plains, their beauty now completely destroyed by a cement works. The heart of the area is Vwawa, a rapidly expanding town of about 10 000 people. There is now a hotel there and sundry other small hostelries. It lies at the edge of the Mbozi 'mbuga', a large stretch of upland flooded grassland rich in flowers. The road leading round the mbuga is called the Mbozi circle road and formerly served the many coffee farms in this area. At Ndolezi is the famous Mbozi meteorite (turn left coming from Mbeya 9 km beyond Ruanda; this is the circle road and 14 km from the turnoff just past Ndolezi on the left is the meteorite. If in doubt ask for 'mawe ya ajabu' or 'kimwondo'). The great mountain of Ngamba (1 877 m) dominates the area. The view from the top is very extensive. West from Mbozi a rather poor road struggles across the plain to the Rukwa (Mkulwe and Kamsamba). Mkulwe is 113 km from Mlowo. This is a fascinating road traversed briefly by Mrs Richards in November. Numerous large mbugas occur along the roadside and are always worth visiting for flowers. At Ivuna on the road to Kamsamba are brine springs now worked as an 'ujamaa' village for the extraction of salt; they are worth visiting. From Kamsamba or Ivuna, Lake Rukwa is accessible and of course is a naturalist's paradise (see again Condry's descriptions).

The Mbeya Range, briefly visited by Mary on her first trip round here (she probably couldn't resist a wee climb!), is a remarkable range with a series of sharp peaks culminating in Mbeya Peak. The whole range is easily accessible from Mbeya town and the Peak is best ascended by Land Rover and then on foot from its north side. The road taken is that to Chunya but at the top near Kawetire Prison (10 km from Mbeya) turn left (westwards) along a rough track that leads to the north side of the peak whence it is 1-2 hours on foot. The view is fantastic if you choose the right day. Another route up the peak for the tough, is from Lunji farm at its south side which is accessible from Mbeya via Mbalizi turning right in the centre of the village, then through Old Mbalizi bearing left for the rest of the way (ask for 'shamba la Bwana Shamshu'). The farm is 8 km from Mbalizi and from it one can take the central gullies, the left one being very severe, or follow paths on the right of the peak.

Another area of interest, briefly visited by Mrs Richards in 1963, is the Songwe valley, access to which is by a very bad road that turns west from the main Vwawa to Mbeya road, 29 km from the latter (just on the Vwawa side of the new Songwe bridge a few kilometres beyond the cement factory). After 8 km this road stops at a lake from where a walk of a few hundred metres brings one to some cliffs and the famous bat caves. If you see anybody then ask for 'pango la popo'. The country here is made up of a limestone formed by deposition of calcium carbonate from springs. It is a botanical paradise and many rare flowers have turned up. From the caves it is a further 1-2 hours walk to the hot springs at Malonde which can be seen from the scarp above the caves as a large white area on the hillside.

Finally, in this Songwe valley region west of Mbeya is the Galula area, the road to which branches off from the Mbeya-Vwawa road at Mbalizi passing through Mchewe (24 km from Mbalizi) and Njurenje and on to Galula. Galula has a church built in the late 20's which is worth seeing. From Galula, a rough track leads to the very shore of Lake Rukwa which at the moment is very high and numerous trees will be seen emerging from the lake. Shortly before the bridge over the Songwe River which leads to Galula is a branch road on the right by which, in the dry season, one can arrive at Mwambani (26 km from the branch). 8 km from the branch is the village of Kanga and a track leads off from here on the right to Lake Masanu, a spot well worth visiting in the wet season when the flamingoes are about.

Mrs Richards entered the Rungwe region (Unyakyusa) for the first time in mid-October 1956 and made numerous collections there. She must have felt much as most of us did first entering this region, that it is the Garden of God ('bustani ya Mungu' as is often said in Swahili). One travels now on a tarmac road from Uyole up to Mporoto or Cimambwe (2 265 m) and 26 km from Mbeya through the heart of the Usafwa region. Then one commences a superb downward run to the large village of Kiwira or Mwankenja (57 km from Mbeya). Several side-trips can be made to points of interest which will be mentioned later in connection with Mrs Richards' visits. Whatever the time of the year, Unyakyusa is green, in sharp contrast to most of Tanzania. This greeness stems from cloud cover and high rainfall which falls all the year round except in September and October. Unfortunately, much of the woodland and forest has been destroyed for tea-planting but, in the areas that remain, the flora is very rich. This was the collecting ground of the German missionary, Adolf Stolz before the first world war (Kyimbila and round about). Typical high rainfall trees here are *Adina,* many kinds of figs, *Myrianthus, Anthocleista, Chlorophora, Raphia* palm and particularly *Trichilia.* Many exotics flourish here such as the Indian teak, Persian lilac, *Toona* and *Ceiba.* It is altogether another world and of course we must not forget to mention millions of bananas of every shape, size and colour!

Places of interest in Unyakyusa are numerous. The Bridge of God ('daraja la Mungu') on the Kiwira River near Kyimo (turn off 6 km past Kiwira on the Kiwira to Tukuyu road and it is 9 km further) is well worth a visit. Here the Kiwira River has undercut a basalt lava flow to form a natural bridge still commonly used by the local people. Higher up the river is the great sump where shortly above the road bridge the river sinks and then surges up again. Rungwe Mountain, a gigantic volcano entirely encircled by a rain forest thickest on the south (wettest) side, can be climbed in about five hours from Rungwe Moravian mission (enquire there for the path which is not easy to find). Another route up the mountain is from the Kiwira Forest Station side on the northeast slope. Follow the Mbeya to Tukuyu road to Isongole, 33 km from Mbeya, turn right at the bridge and follow the forest road round the slope of Rungwe keeping right for 18 km from Isongole. Here, near the end of a pine plantation, a track leads off on the right to the foot of the steep slope of Rungwe and continues up it to a ridge and the crater rim. Descend into the deep hole of the explosion crater and cross towards the main summit ascending the right ridge of the main peak to the top (about two hours from the end of the track). A little before the bridge over the Kiwira a track on the left leads to the famous Kiwira Middle Fishing Camp, still open to visitors, with overnight accommodation available. This is a famous botanical locality and one often visited by Mrs Richards. Another volcano and crater lake often visited in this area is Ngosi. The track turns off the main Isongole to Kiwira road 2 km beyond Isongole towards Kiwira; follow the track for about 3 km towards the foot of Ngosi whence a steep ascent of one hour to the crater rim affords a superb view of the lake. The crater lake of Masoko is also worth visiting; take the Tukuyu to Ipinda road and stop some 18 km from Tukuyu, search on the right just past Masoko village and you should see the lovely lake below you. Another 'must' in Unyakyusa is the waterfall at Makete or Kaparogwe approached on the Makete road which branches off the Tukuyu to Kyela road at Ushirika, 8 km from Tukuyu. At Makete leprosarium, 8 km from Ushirika, ask for Kaparogwe 'maporomoko ya maji'. Unfortunately, the lovely vegetation at the base of the falls where several rare flowers were found has been ruthlessly destroyed. There is also a superb path that leads down the north side of the Kaparogwe valley (a lovely place with prominent *Bombax stolzii* trees and many others) to the Kiwira River where a bridge crosses the river. Matema, 26 km beyond Ipinda, is a holiday centre with an old mission house converted into a guest house. The lake nearby is full of unusual fish and it is a convenient centre for walking and climbing in the Livingstone Mountains. Matema can be booked through Manow (P.O.Box 445, Tukuyu). Kiejo volcano (2 170 m), active about 1800 AD with an eruption from Sarabwe, can easily be climbed by following the Manow road that branches off at Masebe 24 km from Tukuyu on the Tukuyu to Mwakeleli road. On the way, look out for the caves at the base of small cones (full of bats which are both smelly and noisy!). The lava slopes of Kiejo are rich in flowers, one of the most beautiful being *Begonia sutherlandii* growing in the rock clefts. One could mention other areas of great natural beauty and interest for the flower-lover and aesthete such as Itete (hospital) with its crater lake Chungururu; the Kalambo hot springs area; the Mwakaleli area and the forests leading up to Kitulo, Mwaya on the north coast of Lake Nyasa; Ilima and the Lwangwa valley; Ntitu, the big black rock of basalt riding like a ship on softer sediments of the Karroo (accessible with difficulty from the main Kyela-Tukuyu road near the Ilima turn-off); and the lovely woods of Chivanjee and Musekera which alas are rapidly being cut down for tea!

On the west bank of the Kiwira is the equally lovely country of the Bundali or Undali Hills where Adolf Stolz collected from his base at Isoko Mission (1 600 m) on their south flank. The rain forests are rich in tree species and epiphytic orchids. Access by road to this area is very difficult but, in the dry season, one can follow the road from Tukuyu to Ikuti, Katangele (2 100 m) and thence to Isoko. Katangele is a superbly placed village with tremendous views over Lake Nyasa and Malawi. Katangele and Isoko can also be reached by a hazardous road suitable only for four-wheel drive vehicles, from Ilembo in the Umalila via Ibaba (56 km trom Ilembo to Isoko). This is a superb mountain road perhaps the finest in the Southern Highlands (for its scenery of course!) with the grasslands usually full of flowers. The normal approach to Isoko at the moment is from Vwawa via lower Itumba and Malangali. On the way one passes on the left (before Itumba and after Chitete) the great mountain mass of Ileje (2 053 m) best climbed from near Mlale by the road that goes to Mlale from Chitete. There is a beautiful rain forest on its eastern slopes. The Ileje District (Ulambya) and Undali were never visited by Mary Richards in her travels but several flowers new to the East African flora have been found there recently.

On 19 September 1956 Mary Richards left Rungwe and visited one of her favourite and most delectable collecting and camping spots, the Ndumbi River. The Ndumbi River is accessible today from the Kitulo Farm Scheme by a road which branches off left shortly before the bridge on the way leading directly into the living area of the scheme (67 km from Mbeya). To get to Ndumbi, Mary Richards travelled over the Poroto Hills through Igoma (32 km from

Mbeya). Igoma lies at the foot of the Kitulo or Elton Plateau and is itself well up (2 350 m) in the Poroto Hills. From Igoma a picturesque but bad road leads north through Irambo (8 km) to the main road at Nsonyanga. The old Mbeya and Tukuyu to Chimala roads formerly passed this way and the track can still be followed but there are now no bridges. From Igoma the road climbs steeply past Mwamapanza Mountain (2 741 m) to Kikondo (2 450 m, 45 km from Mbeya). A path from here to the south goes steeply down to Lake Dhambwe superbly cradled beneath Wehe Mountain, well worth a visit for its scenery, rich vegetation and excellent swimming. On the left (going north) a road leads to Mwela village (2 600 m, 10 km from Kikondo) which must be one of the highest in Tanzania. Magnificent views are afforded far out over the Usangu Plains and west and northwest over hills and more hills to the far distance. From Mwela one can walk to Ishinga Mountain in one hour (2 687 m) capped by a triangulation stand. The whole country here is rich in flowers from August to May. *Moraea schimperi, Cyrtanthus breviflorus, Kniphofia grantii, Holothrix triloba, Aster tanzanicus, Disa hamatopetala* and *Disa longilabris* flower in vast waves covering the whole country. A track 7 km further on from this turn-off on the left, leads to Matamba and halfway along on the left is a branch to Shansashi (2 450 m), one of the first pyrethrum growing areas in Tanzania. A drop can be made by foot to see the great falls of Salala (Safwa for waterfall) on the Luvaha and Kansasyi Rivers. In a series of cascades and sheer leaps the two rivers fall 450 m towards Igurusi and the Usangu flats.

From Kikondo the road continues round the slope of Chaluhangi Mountain (2 931 m), probably an eruptive vent, which is easily climbed, giving superb views over Unyakyusa. The Kitulo Farm Scheme is soon reached and the whole scene is reminiscent of Europe rather than Africa (at least it certainly was when there were 14 000 sheep there). The turn off to Ndumbi is just before the scheme and this is the same road that leads to Tandala in Ukinga. Another road to Bulongwa (very hectic) turns off the Kitulo road a few kilometres from the scheme on the way back to Mbeya and is exciting, floriferous and scenic. Kitulo itself is 67 km from Mbeya; Tandala is 77 km from Kitulo via Kisinga and is a very interesting road for flowers and scenery. Look in the swamps for orchids and gladioli and many other plants. Following the Tandala road from Kitulo for 8 km one can branch off left up the ridges of Mtorwi Mountain (2 960 m) for 2 km, leave the transport then climb the highest mountain in Southern Tanzania. Mary Richards climbed it on 13 January 1957 and collected many species, the shrubby *Philippia pallidiflora, Helichrysum* by the dozen (*globosum, latifolium,* etc.), *Berkheya echinacea,* the lovely little shrub *Hebenstreitia dentata, Crassula alba* everywhere, *Leonotis leonitis* with its conspicuous large red flowers, and countless *Cephalaria pungens.* There is always something in flower on this mountain and it is rich in orchids in the wet season, particularly on and around its summit.

The Kitulo or Elton Plateau is of course famous in botanical literature and the flowers just have to be seen to be believed; imagine in November with the rains beginning kilometre after kilometre of rolling uplands at 2 500 m covered by millions and millions of *Aster tanzanicus, Kniphofia grantii* (one to every square metre at least!), *Moraea schimperi, Pentas decora, Berkheya* and *Geranium* species, *Scutellaria pauciflora, Becium cameronii, Scilla rigidifolia, Dierama pendulum, Justicia nuttii, Anthericum, Helichrysum* and *Cephalaria* species and innumerable legumes. Later, in December, the orchids start to appear and go on until June. The plateau, a unique world needing a book to itself, is a must for any flower lover and the bird life is also of great interest. Mrs Richards was up here often from 1956 onwards (January 1957 and 1961 and November 1963). A favourite area of hers, slightly off the plateau, was Matamba (2 132 m) in the heartland of the Wanji people, a superb area at high elevation with the backdrop of the high tops of the Kitulo Plateau. The road to Matamba branches off the Mbeya-Chimala main road 3 km from Chimala on the Mbeya side (signpost) and climbs some 55 bends (try and check them!) to the top of the escarpment. Flowers are plentiful there throughout the wet season and the flushes of the *Brachystegia* trees in October and November are famous (see descriptions in Condry's book). Past Matamba, the road continues over the Matamba Pass to join up with the Mbeya to Kitulo road (18 km from Matamba). Do not fail to look on the rocky ground by the pass for flowers.

A final section of the Mbeya Highlands is the country between Chunya and Gua, north of Lake Rukwa. The road into this area has long been praised as a scenic highway (highway it certainly is not, scenic yes!). From Mbeya, the road branches off just outside the town past the Guest House and climbs to near the prison at Kawetire, then on past a sign claiming 'highest point of any main road in Tanzania' (2 459 m) to Chalangwa village and Chunya (72 km). Mrs Richards passed this way in December 1963 and continued down to Bangala on the lake where there was a hotel which has since gone under water (the lake has been rising since 1955). At Saza, an old mining village, one can branch off to Mkwajuni and Mwambani and in the dry season return to Mbeya (Mwambani to Mbeya 102 km). From Saza, the road struggles on to Gua. At Kwimba (38 km from Saza) the car can be left and Kwimba Mt can be climbed (1 694 m) in two hours or less; plenty of game here and in the wet season trees and flowers in abundance. On the top is a great rock and the whole mountain is sacred to the Wabungu tribe of this region and many others. Although really out of the Southern Highlands a road goes north from Makongolosi to Itigi passing Kipembawe (99 km from Makongolosi) and Mary Richards collected up this road in March 1965.

3. *The Njombe Highlands*

These include the greatest area of high ground in the Southern Highlands of which three sections can be distinguished. There is the rather monotonous plateau of Ubena at 1 800 m, falling north to the Usangu Flats on the southwest and broken by the NW-SE trending higher ridges of the Ubendian and Ukingan rocks, continuous with the Kitulo plateau. The Gofio plateau and the superb escarpment of the Kipengerasi above the Mbarali River are formed of gently dipping Buanji Sandstones. The third portion is the so-called Livingstone Mountains.

Mrs Richards made a few excursions in this area. She often used the name 'Kipengere' for localities and it must be taken to mean the highlands around Kisinga and the southeast end of the Kitulo Plateau. She was there in January 1957. She collected at the Rumakali River and Mwakete and from Ukinga she went over the hills to Njombe. This road from Bulongwa via Tandala to Njombe is hectic in the wet season but is a superb scenic treat. Anybody going to Mbeya from Iringa who trusts his transport could make this detour. The road branches at Makambako to Njombe (61 km) a pleasant, breezy town at 1 850 m. Before entering the town, the road branches right for Kipengere (36 km from Njombe) and then onwards over the tops and bottoms to Tandala, an attractive run of 40 km. From Tandala, one can go to Bulongwa and from there continue out by road to the edge of the plateau and look down on Lake Nyasa or even drop down by foot, an exhilarating trip, to Matema. The forest has giant 'mvule' (*Chlorophora excelsa*) trees as well as great groves of tree ferns. From Tandala (an excellent centre for botanising) the road also goes on to Kitulo via Kisinga. A short branch goes to Ukwama, the heartland of the Wakinga people (15 km).

In January 1961, Mary Richards made one of her two safaris into Upangwa and passed through the heart of the second section of the Njombe Highlands. The road goes south from Njombe (there is a short cut to Uwemba, 16 km from Njombe, branching off near the Anglican cathedral in Njombe) then south over the high plain until one reaches the mountains at Matola (56 km from Njombe), lovely undulating country with superb swamps rich in flowers. 82 km from Njombe there is a colossal escarpment and the road drops down into Upangwa and out of Ubena. In the distance the Livingstone Mountains ride on the horizon looking for all the world like an African Himalayas, their sharp peaks contrasting with the rounded heights of Ubena and the rest of Upangwa. Mary went on to Milo, the old Anglican (UMCA) mission in the heart of Upangwa. The road passes through Mlangali (104 km from Njombe) and then up a steep hill and winds along the top of a plateau to Milo (2 085 m above sea level) which is a beautifully situated station on a crest of the hills above the Luimo River. From Milo, she went in early February to Madunda, an excellent centre from which to explore the Livingstone Mountains. We do not seem to have any local name for this range of mountains which stretches down the northeastern side of Lake Nyasa for some 60 km. Most of the main mountains have local names such as Idofu, Nyamlunga (probably the highest at c.2 450 m), Shoga, Lipatu, Liholo, Ligala (above Madunda), Masusa (above Kilondo on the Lake shore) and Kimulimuli (above Lumbila). Many of the

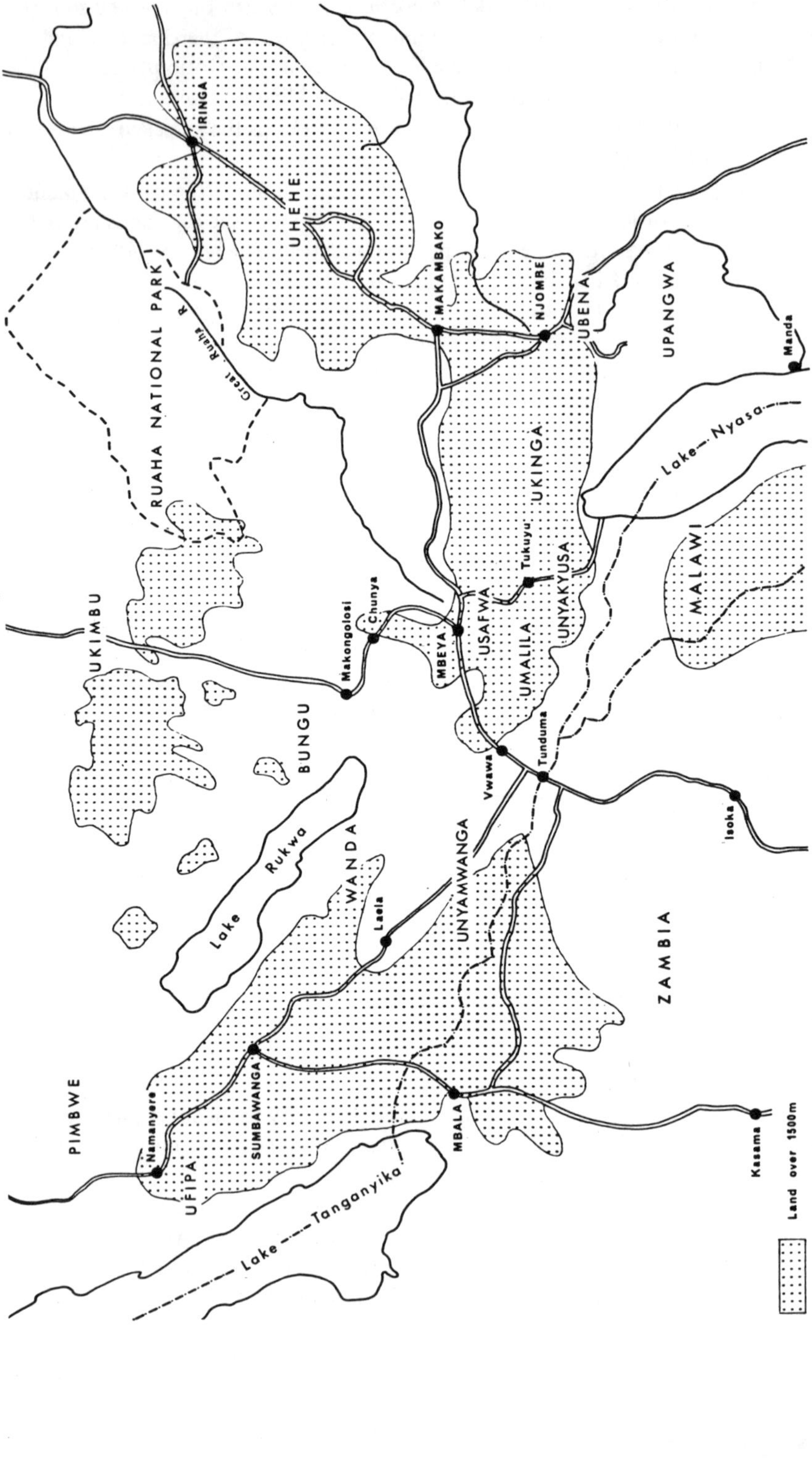

Figure 1. The Southern Highlands of Tanzania

peaks have not been climbed but most of the main peaks should be accessible to any strong walker. Ligala with its conspicuous cross erected near the top by the local mission is readily climbed from Madunda Mission in about one and a half hours and affords superb views over Lake Nyasa across to the Nyika Plateau in Malawi. Patches of rain forest occur on most of the Livingstone summits. From Madunda the road winds southwards to Manda through Rudewa district; this is a scenic trip but the road is bad. The Ruhuhu River could be conveniently counted as the border of the Southern Highlands and further south the high ground can be referred to as the Songea Highlands.

A final section of the Njombe Highlands which must be mentioned is the Kimani River, Ruaha River and Kipengerasi Escarpment. This country lying to the south of the main road from Makambako to Chimala is a superb stretch of country, well worth visiting for its flowers. One crosses the Great Ruaha River a few kilometres after leaving Chimala and, a short distance further on, a turning on the right leads through the valley of the Great Ruaha up on to the Ikuwo plateau area beneath the peak of Mtorwi. Five kilometres past the Great Ruaha bridge on the main Mbeya to Iringa road, an old road leads to the Kimani River where there is excellent swimming both above and below the old bridge. A turn-off on the right before the old bridge leads for 1 km up the valley where the car can be left. A foot-slog up the valley leads to the Nyalugenge Falls, one of the great sights of Tanzania when in flood at which time they are audible for a long distance. When we were there in mid-June, the giant steps of the Falls were covered with the large-flowered *Gladiolus decoratus.* It must have been under water for two or three months before that. Above them around a sharp bend in the river is a natural bridge where the river sinks into the sandstones. Ask the local people how to get there or follow people going up to Ikuwo. The Kipengerasi Escarpment is best approached from Kidugala. From Kidugala it is 14.5 km to Wangama village and another 8 km up into the hills from which the escarpment is accessible; superb flower country and no doubt a few new species are still to be found there. Between Kimani and the Great Ruaha valleys is the flat-topped mountain of Chafukwa, formed of lava cliffs rich in agates which can be collected at the base. Chafukwa can be readily climbed from Kimani or the main road.

4. *The Mufindi and Dabaga Highlands*

These lie in the easternmost portion of the Southern Highlands forming an enormous expanse of only moderatcly dissected plateau country, at its highest in the east and then dropping abruptly to the Kilombero valley. The highest point is Chuge at 2 445 m in the Dabaga Highlands.

Mrs Richards first came here to collect in December 1961, when she climbed Image Mountain, lying well to the north of Iringa, and then came south to Sao Hill (1 840 m) the heartland of Mufindi. From Sao Hill she

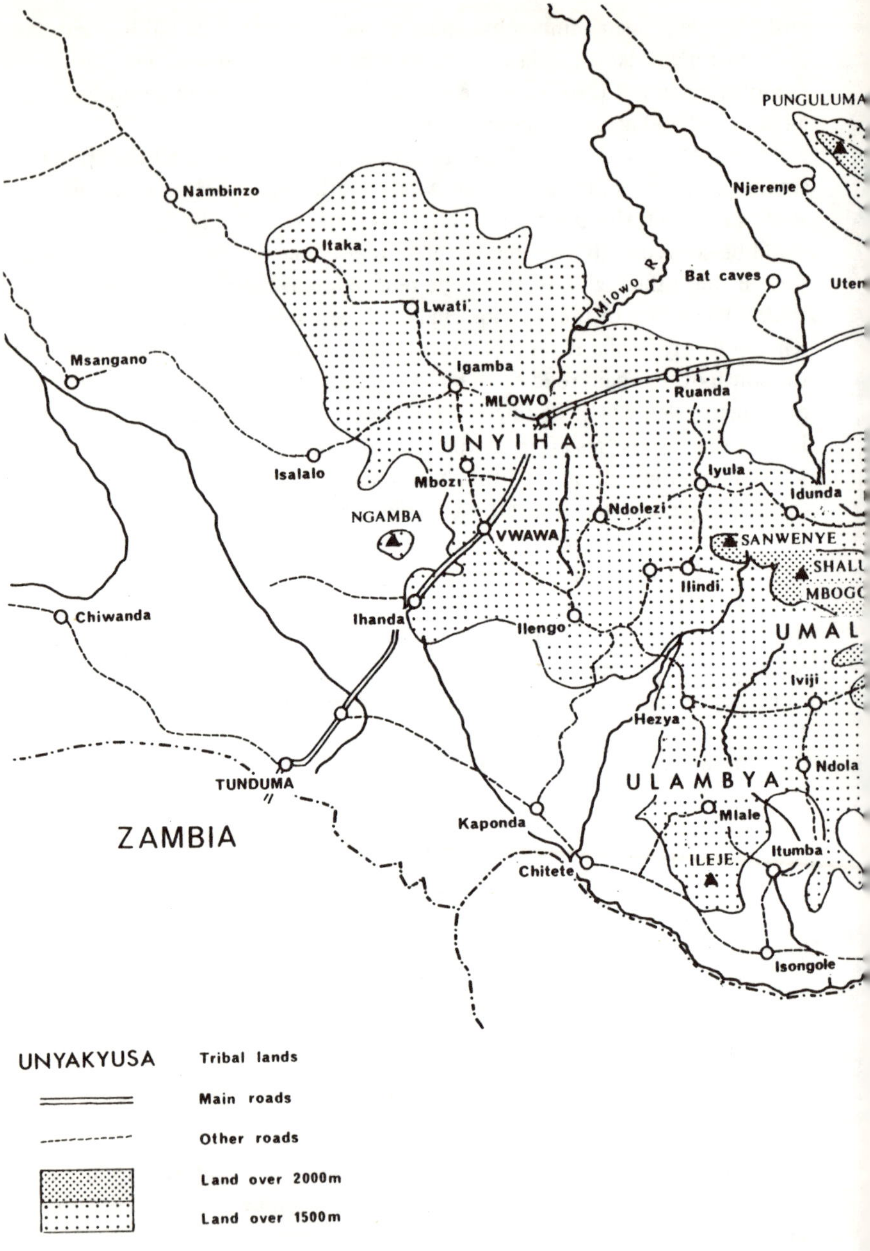

Figure 2. The central area of the Southern Highlands around Mbeya

CHALANGWA
Utengule
Mswiswi
Igurusi
MBEYA PEAK
Inyala
CHIMALA
MBEYA
Uyole
USAFWA
Irambo
Mbalizi
SALALA
Mporoto
Chimala R
NGOZI
Izyonje
Igoma
Matamba
UPOROTO
UANJI
MTORWI
Idweli
Isangati
Kikondo
Ndaga
Kitulo
CHALUHANGI
Santilya
RUNGWE
Kandete
Kiwira
Mwakaleli
Masebe
Katumba
Suma
KIEJO
Ngulilo
TUKUYU
Manow
Lwangwa
UNYAKYUSA
Itete
Imadahani
Ikuti
Masoko
Mbambo
Bulongwa
Chivanjee
Mbaka R
Lufirio R
KISIBA
Kiwira R
Makwale
Katangele
Isoko
Ipinda
Matema
Upigu
Ikikota
KYELA
Itungi
Ilondo
Nyasa Songwe
Lumbila
MALAWI

journeyed to the Kibao Highlands surrounding the tea estates of Brooke-Bond Liebig. On 18 December she visited the Kigogo River and forest and, amongst other species, collected the beautiful shrub *Bersama rosea.*

The Kibao Highlands are easily approached by good roads from the south (Makambako) via Nyolola (48 km from Makambako) where one branches right through Igowole to the tea estates. Superb rivers drain the forest down the 800 m high Mufindi scarp. The forest is magnificent with numerous species such as the giant indigenous African papaw tree *Cylicomorpha parviflora.* The largest river of the area is the Kigogo and the upper reaches are readily accessible. At Luisenga there is a fishing camp on the shores of a dam near the tea factory with a guest house and adjacent golf course. The whole area is full of interest to the naturalist. From the north, Kibao is reached via Mufindi or John's Corner (80 km from Iringa). A worthwhile trip from Kibao is to Mdabulo Mission in the northeast corner of Mufindi (34 km).

The Ruaha National Park visited by Mary Richards in December 1962 and November 1966 lies on the extreme western side of the Mufindi Highlands. A full description of the vegetation of the park (1 650 species) has recently been given by Bjørnstad (1976). The park is accessible (bad road) from Iringa and accommodation is available. No visitor has been displeased with this park and any botanist wishing to see savanna and woodland species should visit it. A duplicate set of Bjørnstad's herbarium specimens is deposited at the Park Headquarters and may be consulted there.

Finally we come to the Dabaga Highlands where there are no outstanding peaks but only a few higher tops on the plateau. Like the Mufindi Highlands there are some fine stretches of forest with typical tree species such as *Bersama, Hagenia, Milletia, Nuxia, Entandrophragma* and many others. Epiphytic orchids flower there from October onwards. Access is from Iringa to Kilolo (39 km) and Dabaga lies 9 km further on. At Isimila, south of Iringa there is a stone age site which is well worth visiting. Iringa town is a good centre to stay in but has little of botanical interest although the orchid *Cyrtorchis crassifolia* should be looked for growing on the *Eucalyptus* trees in the main street.

FAMILIES AND SPECIES

RANUNCULACEAE
THE BUTTERCUP FAMILY

Plates 1, 2 and 3

Commonly herbs or, less frequently, woody climbers such as *Clematis,* the leaves often deeply divided. In *Clematis* the leaf stalks are sensitive and support the stem-like tendrils. Stipules are present in a few genera only such as *Ranunculus* and *Thalictrum.* Inflorescence terminal, the flowers bisexual with their parts commonly arranged in a spiral along a more or less elongate receptacle. Often the sepals are petaloid but occasionally a distinct calyx and corolla are present as in *Ranunculus.* The inner fertile parts of the flower comprise spirals of many stamens and, commonly, many carpels. Both regular flowers as in *Ranunculus* and *Clematis* and bilaterally symmetrical flowers as in *Delphinium* are found in the family. Most species are insect-pollinated but *Thalictrum* is an exception being wind-pollinated.

The family is predominantly a northern hemisphere one but extends southwards into Africa on the higher mountains. The African species like their northern relatives are often decorative plants especially the species of *Delphinium* and *Clematopsis.*

Clematis simensis Fresen. [1A]

This attractive climbing shrubby plant is widespread in the Southern Highlands. It often reaches 20 m or more and becomes a strong liane. The younger shoots are more or less hairy but eventually become glabrous. The leaves are commonly imperfectly 5-pinnate, the leaflets varying from glabrous to hairy. Its inflorescences contain many attractive cream or white flowers 1.5 to 3 cm in diameter.

It is common in the Southern Highlands around Mbeya and in the Ukinga Mts and across to Mufindi where it can be seen in flower in August. Usually found growing on forest margins or in bushland above 1 000 m. The Wanyakyusa name for this species is 'kanyantitu' whist, in Mufindi, the Wahehe called it 'likalichuma'.

Of the six other East African *Clematis* species, two are recorded from the Southern Highlands. *C.commutata* O.Ktze. (syn. *C.iringaensis* Engl.) growing in both the Iringa and Mbeya Highlands is distinguished by its very hairy leaves and the yellow indumentum on the outer surface of the flowers. The other species found here is *C.welwitschii* O.Ktze. which is characterised by its prostrate habit and oblong leaflets. It is called 'mukulasunime' by the Sangu and Nyamwezi people, to whom its flowering at the end of May denotes the end of the rains. It is found over vast tracts of woodland in Ukimbu, Usangu and Usafwa.

Plate 1. A, *Clematis simensis* (A1, fruiting head); B, *Clematopsis villosa*

Clematopsis villosa (DC.) Hutch. (syn. *C.scabiosifolia* (DC.) Hutch.)
[ID & 1B]

One of the most beautiful and ubiquitous of all the plants of the Southern Highlands is the delightful *C.villosa.*

It commonly forms a broad bushy plant up to 1.5 m tall with few to many erect or spreading hairy stems. Its leaves are also softly hairy and very variable in shape from entire to trifoliate or bipinnate. Its attractive nodding flowers are typical of those of the climbing *Clematis* species but large, up to 7 cm across and hairy on their outer surface. The seeds also distinctively have a tuft of grey hairs at their apex.

Several subspecies of *C.villosa* are recognised now in Africa and two of these occur in the Southern Highlands. The most widespread of these is subsp. *kirkii* (Oliv.) J.Raynal & Brummitt which is widespread in the woodlands of south-central Africa from Mozambique and western Tanzania across to Zambia and Angola and south to Zimbabwe. It is readily distinguished by its pinnate leaves with up to three pairs of rather broad leaflets and an odd terminal one. It has branched inflorescences with many sweetly scented creamy white or pale pink flowers. This delightful subspecies is widespread between 1 000 and 2 000 m. It flowers between January and March and can be seen, for example, in the Poroto Mts in the grassland and scrub around Isongole village and also commonly around Mbozi west of Mbeya. The Wamalila people call it 'ikuwi' and the Hehe, 'makalichoma'.

The second subspecies in our area is subsp. *uhehensis* (Engl.) J.Raynal & Brummitt. The Wakinga people who live high up on the Kitulo Plateau call this flower 'masusuko' and the Safwa, 'fwefwefwe'. It is a magnificent sight when it comes into flower in its thousands on the plateau in January in association with pink and purple *Geranium,* white *Pentas decora* and the yellow and reds and pinks of various *Impatiens* species. Subsp. *uhehensis* is distinguished by its simple leaves and solitary flowers which are commonly white and mauvish or rose-purple in colour. It is apparently confined to the higher grasslands of the area around Iringa and Mbeya and is also found in neighbouring Nyika Plateau of Malawi and Zambia.

A second species *C.chrysocarpa* (Welw. ex Oliv.) Hutch. is also found in the Southern Highlands and in neighbouring Malawi, Zambia and Angola. Like *C. villosa* subsp. *uhehensis,* it has solitary flowers but its leaves are normally twice pinnate with an odd terminal leaflet, the leaflets are narrow and the seeds have a golden-brown hairy apex.

Delphinium dasycaulon Fresen. [2A]

Blue flowers are uncommon and always attractive. *D.dasycaulon* commonly has them although its flower colour can vary to mauve, pink or even white. It is an elegant erect herb up to 1 m tall with yellowish hairy stems and inflorescence branches. The deeply dissected leaves are mostly basal and often persist

Plate 2. A, *Delphinium dasycaulon;* B, *D.leroyi;* C, *Thalictrum rhynchocarpum* (C1, flower)

after the flowers have died. The flowers are 2 to 2.5 cm across and have a 5 to 8 mm long stout straight spur.

It is a widespread but uncommon species in our area mostly found in the woodlands between 1 700 and 1 800 m. It has been recorded from the Ufipa area across to the Njombe Highlands and has even been collected on the top of Mbeya Mt. Mrs Richards found it on Kito Mt. in south Ufipa in late April growing in association with several legumes such as *Astragulus atropilosulus* and *Lotus discolor* and other pretty species including the milkwort *Polygala petitiana.*

Elsewhere in Africa *D.dasycaulon* is widely distributed from Ethiopia and the Sudan as far south as Malawi and Zambia and across to Cameroun in the west.

Delphinium leroyi Franch. ex Huth [2B]

This, the second species of *Delphinium* found in the Southern Highlands, is another beautiful plant. It is a large erect perennial herb from 0.5 to 1.5 m tall with pubescent stems and dissected leaves borne on long stalks at the base of the plant. The sweetly scented flowers are unmistakable, up to 6 cm across, white with sometimes a green spot on each segment and a long ascending spur, 3 to 4 cm long, from the back.

D.leroyi is widespread in East and Central Africa from the Southern Sudan and Ethiopia south to Malawi growing in montane grassland, on forest margins and in bushland from 1 300 to 2 900 m. It can, however, be seen at its best in the montane grasslands of the Poroto Mts and Kitulo Plateau where it flowers from February until May often in association with the delightful yellow-flowered orchid *Satyrium sphaeranthum.* In the Ukinga country where it is widespread it is called 'manyambimbwi'.

Thalictrum rhynchocarpum Dillon & A.Rich. [2C]

Only four species of the predominantly north temperate genus *Thalictrum* have been recorded from East Africa and three of these are found in our area.

The commonest of these is *T.rhynchocarpum,* a widespread plant throughout tropical and southern Africa. It is a perennial herb, 1 to 4 m tall, with erect more or less branching stems. Its delicate leaves, which reach 40 cm long, are highly divided and remind one of those of the maidenhair fern (*Adiantum*) so much so that many people are fooled when they first find this plant. The inflorescence is laxly many-flowered. The flowers are up to 1.7 cm long with short green or purplish sepals up to 4 mm long only. Most of the flowers are bisexual but some are male only.

Mrs Richards collected this species many times but never failed to be charmed anew each time she saw it. One of her collections was made on the Nyerere River area on the Kitulo Plateau in January 1957. The plant grows there in association with the bright pink marsh orchid *Satyrium crassicaule,*

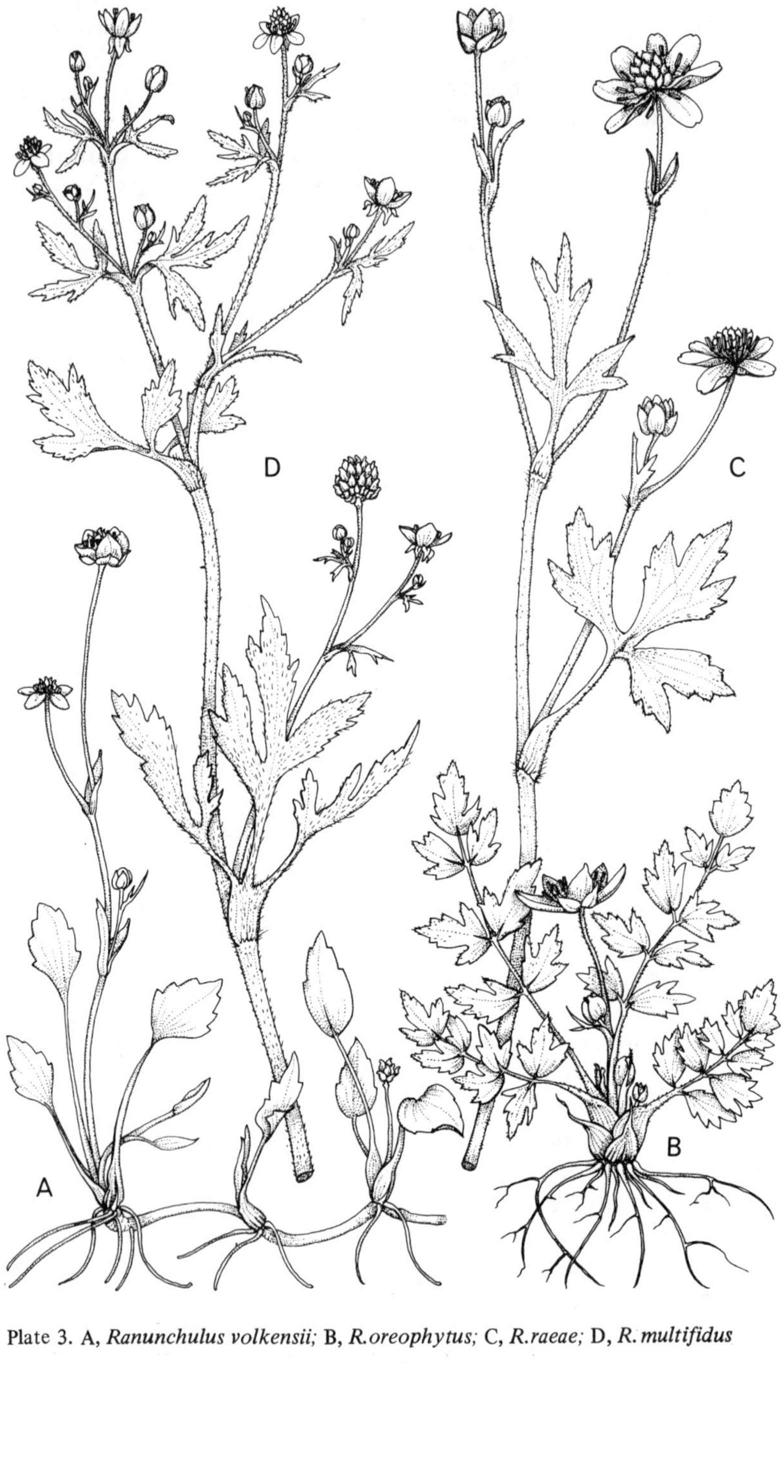

Plate 3. A, *Ranunchulus volkensii;* B, *R.oreophytus;* C, *R.raeae;* D, *R. multifidus*

the sage *Salvia nilotica* and the milkweed *Ceropegia chrysocoma.* It occurs in shady places and in forest throughout the Southern Highlands above about 1 600 m. The Safwa people call it 'zeleleho' and use its roots medicinally.

The two other species of *Thalictrum* recorded hereabouts are *T.stolzii* Ulbr. found in bamboo thickets near Njombe, a rare endemic distinguished by its compact few-flowered inflorescence, and *T.zernyi* Ulbr. The latter is a lovely plant also confined to the Southern Highlands area. Its flowers are pale rose-pink each producing 10 to 14 seeds. It is a small plant from 15 to 50 cm tall found growing on steep rocky ground or on cliffs in the hills around the top end of Lake Nyasa at between 1 800 and 2 800 m.

Ranunculus multifidus Forsk. [3D]

Nine species of buttercup have been recorded in East Africa but only four of them are found in our area.

R.multifidus is probably the most widespread of these ranging as it does from Arabia south to the Cape and west to Nigeria. It is also a very common plant of wet grasslands above about 1 200 m in the Southern Highlands. It is a perennial herb with erect or more or less prostrate hairy stems rooting at the nodes. Its leaves are very variable in their lobing and rather variably hairy as well. The small yellow buttercup flowers are 12 to 16 mm across with reflexed sepals and are borne on the upper part of the much branched stems. It can be seen in flower from December until February. Mrs Richards collected the plant illustrated here at Kikondo on the edge of the Kitulo Plateai where it grows with *Lotus goetzei, Indigofera mimosoides, Adenocarpus mannii, Crepis kilimandscharica, Psoralea foliosa, Buchnera rungwensis* and another buttercup *R.volkensii.*

Ranunculus oreophytus Del. [3B]

R.oreophytus is another widespread buttercup in East Africa but is essentially a species of the high grassland above 2 600 m, an element of the so-called Afro-alpine flora.

It is a stemless perennial herb with a rosette of 2- to 4-pinnate, glossy, dark green leaves, the leaflets being coarsely toothed. The yellow flowers, borne singly on stalks up to 10 cm long, are about 2 to 4 cm across. Of particular interest is the habit of the fruiting heads which curve over and bury themselves in the soil as the achenes mature.

R.oreophytus is common on the Kitulo Plateau. Although the pasture improvements adopted by Kitulo Farm Scheme have changed the face of much of the formerly flower-strewn pastures of the area, a great deal still remains to be seen there. From 1967 until 1969 Mr and Mrs Prins, who were attached to the scheme, collected over 270 species in the grasslands on the farm and these form the basis of a small herbarium kept there at present.

Around Kitulo, *R.oreophytus* commonly grows with its rare white- or purplish-flowered relative *Knowltonia transvaalensis* as well as the allied beauty *Delphinium leroyi.*

Ranunculus raeae Exell [3C]

Mrs Richards collected this buttercup many times on her travels and it is a widespread species in the Southern Highlands between 1 300 and 2 600 m.

R.raeae is a tuberous rooted perennial herb up to 60 cm tall with palmately lobed leaves about 4-5 cm across. It has small yellow flowers 1.5 to 2 cm across. The achenes have a straight or slightly curved beak and are borne in a spherical head.

In the Southern Highlands it commonly grows in association with the curious little flower *Psammatropha myriantha* (Aizoaceae) which is restricted to this area in East Africa. *Thesium kilimandscharicum,* the curious shrub *Phylica emirnensis, Hedythyrsus thamnoideus* and *Myrsine africana* also grow in this association.

R.raeae is confined to western and southern Tanzania in East Africa and is also recorded from Zaire and Zambia.

Ranunculus volkensii Engl. [3A]

This is yet another buttercup of the Kitulo Plateau and was collected there several times by Mrs Richards, growing on high ground above 2 300 m and flowering from October until December.

R.volkensii is a creeping perennial herb with slender prostrate stems rooting at the nodes. The leaves are characteristically simple with either an entire or toothed margin. The small 8 to 14 mm diameter yellow flowers are borne singly at the nodes on 1 to 8.5 cm long stalks.

It grows in association with a variety of colourful flowers such as the yellow-flowered *Crepis iringanus* and *Hypericum scioanum,* the purple-flowered *Cynorkis anacamptoides* and the white trumpets of *Pentas decora.*

R.volkensii was originally collected by Volkens on Mt Kilimanjaro and it is widespread in East Africa and also in Zaire, Rwanda and Burundi at altitudes between 2 700 and 4 500 m.

CARYOPHYLLACEAE
THE PINK OR CARNATION FAMILY

Plate 4A and B

Usually annual or perennial herbs with opposite, simple and entire leaves. Stems commonly swollen at the lower joints and stipules usually absent. Inflorescences cymose of regular and bisexual flowers. Calyx consisting of four or five free or united sepals and corolla of a similar number of free petals often sharply divided into a basal claw and limb and often lobed or fringed at the apex. Stamens commonly twice as many as petals and usually attached to the receptacle. Ovary superior with two to five free styles. Fruit usually a capsule dehiscing at its apex.

This is another predominantly north-temperate family, most of its African species confined to the higher mountain areas. *Silene,* the catchflies or campions, and *Cerastium,* the chickweeds, are the two commonest genera in our area.

Cerastium afromontanum T.Fries & Weimarck [4A]

Three species of this predominantly north-temperate genus occur in East Africa and all three of these have been recorded from the Southern Highlands.

C.afromontanum is the most restricted of these and is a more or less glandular hairy slender creeping or ascending plant with stems up to 40 cm long. The leaves are from 4 to 35 mm long and 2 to 9 mm broad. The inflorescences are terminal and bear from 2 to 23 flowers. The petals are 8 to 11 mm long and the sepals are glandular hairy.

It is widespread in East Africa from Sudan south to our area and grows in upland grassland and moorland over 2 100 m altitude. In the Mbeya region it can be found between 2 300 and 2 700 m, growing amongst grasses in shade in the Poroto and Kipengere Mts and on the north slopes of Mt Rungwe amongst thickish vegetation. It flowers thereabouts from December until February.

Silene burchellii Otth. ex DC. [4B]

Silene is another predominantly European genus with only a few species, in this case three, in East Africa. Of these, *S.burchellii* is the most widespread and attractive.

It is a perennial herb, 30 to 70 cm tall, growing from more or less tuberous roots which are claimed to be edible. The stems, leaves and calyx are scabrid and the leaves range from 1.8 to 9 cm long and 0.4 to 1.4 cm broad. Its inflorescences are quite distinctive with all the flowers pointing to one side. The

Plate 4. A, *Cerastium afromontanum;* B, *Silene burchellii;* C, *Linum volkensii*

flowers are 1.4 to 2.2 cm long with a pale green calyx lined with purple-brown and with creamy or pinkish corolla lobes up to 6.5 mm long.

Commonly it is found growing in open woodland, in stony places, on screes and in upland grasslands over 1 500 m. It is widespread in the Southern Highlands and can be seen flowering from December until March on the Chunya and Chimala escarpments in grassy places in woodland and around Mbozi, west of Mbeya.

This species is found throughout Africa from South Africa to Eritrea and Ethiopia and also in the Yemen.

LINACEAE
THE FLAX FAMILY

Herbs or shrubs with usually alternate, small, entire leaves with or without stipules. Flowers regular and bisexual with parts usually in fives. Petals free, generally withering and falling quickly. Stamens equal petals in number, are alternate with them and have the base of the filaments fused in a ring. The ovary is superior and generally gives rise to a capsule.

Linum usitatissimum, flax, produces a fibre from its stems which can be woven into linen and also linseed oil from its pressed seed. The latter is used in the manufacture of paints, varnishes and printing ink.

Linum volkensii Engl. (syn. *L.holstii* Wilczek) [4C]
This dainty flax is a common plant of the marshes and grasslands of the Southern Highlands. It stands up to 90 cm tall and its slender, alternate leaves range from 5 to 25 mm long and 1 to 3 mm broad. The flowers are yellow or orange-yellow with a green calyx and the petals are 6 to 9 mm long.

It has been recorded from Mufindi where it grows in grassland on the Lugoda Tea Estate across to the Mbeya range and the Kiejo volcano where it grows in thin soil over lava amongst scattered bushes. It flowers in our area from March until September.

L.volkensii is widely distributed in the eastern half of Africa from southern Ethiopia southwards to Mozambique and Malawi.

Two related species, *L.thunbergii* Eckl. & Zeyh. and *L.keniense* T.Fries, both commonly yellow-flowered, also grow in our area. The former differs in having opposite lower leaves and sepals with a prominent midrib, whilst the latter differs in having whorls of three or four leaves which are rounded at the apex.

GUTTIFERAE
THE ST JOHN'S WORT FAMILY

Plate 5

Herbs, shrubs or trees with opposite entire leaves, often gland-dotted in *Hypericum.* Flowers bisexual in African species, the parts borne in whorls of four or five, usually with distinct sepals and colourful petals. Stamens many, born in two whorls, the outer often sterile, around a superior ovary which has three to five free styles at its apex.

The most obvious African genus of this cosmopolitan family is *Hypericum* popularly known as St John's Worts.

Hypericum lalandii Choisy [5A]

The genus *Hypericum* contains a large number of species including some 15 in East Africa. These exhibit a great variety of form ranging from small creeping perennial herbs to small trees yet all are readily recognised by their distinctive flowers, quadrangular young stems and often gland-dotted leaves.

H.lalandii is one of the smaller species, a tufted perennial herb up to 70 cm tall with 4-angled stems somewhat woody below and gland-dotted leaves, 1-2.2 cm long. Up to 50 flowers are borne in a lax inflorescence. These open orange or yellow from a red bud and are commonly 1.2-1.6 cm across. The fruits are dark red capsules.

This attractive species is common throughout Africa from Sudan to the Cape and is also recorded from Madagascar and mainland Asia. In Africa it commonly grows in marshes and wet grassland between 1 080 and 2 250 m altitude. It is common on the north slopes of the Poroto Mts and can be seen near Mporoto at between 2 000 and 2 400 m growing among grasses and bracken and flowering in April and May.

Hypericum lanceolatum Lam. [5B]

As widespread in distribution as *H.quartinianum,* this species almost matches it in flower size. Commonly *H.lanceolatum* forms a shrub or even small tree with scaly bark from 1 to 12 m tall. Its small bright green leaves, 1.2-2.2 cm long, are borne on the young 4-angled stem. The leaves have translucent surface and marginal glands. The solitary flowers are bright yellow or orange and up to 5 cm across. The sepals are green with black marginal dots.

H.lanceolatum is found on most of the mountains of Tanzania growing in dry evergreen forest, in upland bushland and by streams in upland grassland between 1 800 and 3 360 m. In our area we have seen it on the Chunya escarp-

Plate 5. A, *Hypericum lalandii;* B, *H.lanceolatum;* C, *H.quartinianum;* D, *H.scioanum*

ment north of Mbeya growing by a roadside at 2 340 m and in the ericaceous scrub near the top of Mt Rungwe at 2 910 m. In the Southern Highlands, it flowers from July until December.

Hypericum quartinianum A.Rich. [5C]
This attractive shrub is almost at the southernmost extent of its range in the Southern Highlands for it is widely distributed in East Africa from Ethiopia south to northern Malawi.

It reaches 4.5 m tall, the lower stems being woody and tawny-brown while the young shoots are distinctly 4-angled. The leaves reach 8 x 2.2 cm and when young are tinged with red. These have black dots on their margin and glands on their upper surface. The flowers, reminiscent of those of the cultivated Rose of Sharon, are 5 to 8 cm across and butter-yellow with pale mauve stigmas.

It grows in rocky places, in gulleys and on river banks in deciduous woodland or in upland grassland. On the Kitulo Plateau it can be seen at altitudes up to 2 800 m, by the Iringa to Mbeya road at Ilango at 1 800 m and at the foot of Ngosi Mt in August at 2 200 m. It flowers from May until September in our area.

This species and *H.lanceolatum* are called 'lusaladanda' by the Uanji, 'linunadanga' by the Wapangwa and 'lufadanda' and 'lufwadanga' by the Kinga.

Hypericum scioanum Chiov. [5D]
Although originally described from a plant collected in Ethiopia, this diminutive species is widespread in the mountains of East Africa and in Zaire.

The creeping or ascending shoots reach 30 cm long and bear small gland-dotted leaves up to 9 mm long but plants from higher altitudes are generally much smaller in all their parts. Its yellow flowers are borne singly and are commonly 8-10 mm in diameter.

H.scioanum is widespread in the mountains of Tanzania and may be seen throughout our area from 2 200 to 2 960 m altitude. It usually grows in boggy places by streams and in grassland such as by the track to the Ngozi Crater and by the Kiwira fishing camp on Mt Rungwe. It flowers there from October until December.

Plate 6. A, *Abutilon longicuspe;* B, *Pavonia urens*

MALVACEAE
THE MALLOW FAMILY

Herbs, shrubs or trees with alternate often stellately hairy leaves. Flowers regular, bisexual, with parts in fives; the sepals sometimes joined, the petals free but the stamens united below into a tube around the superior ovary. Fruit commonly a dry capsule.

This cosmopolitan family includes the attractive garden species of *Hibiscus* and *Abutilon* as well as the popular European garden plant, *Althea rosea,* the hollyhock. The genus of the greatest economic importance is *Gossypium,* cotton, and *Hibiscus esculentus,* okra, is widely eaten.

Abutilon longicuspe Hochst. ex A.Rich. [6A]

A scandent perennial shrub with pubescent stems 1 to 5 m tall. Its leaves, about 8 to 19 cm long and 7 to 12 cm broad, are stellately pubescent and borne on stalks 8 to 10 cm long. The flowers, 2.5 to 3.6 cm in diameter, are pale lilac to mauve with a red blotch at the base and radiating darker venation on each petal, the central filaments reddish-purple, about 1.5 cm long. Its fruits are covered with stellate hairs.

It is a widespread plant of lower montane areas from 1 000 to 2 100 m throughout tropical East and South Africa, growing in secondary forest and flowering from July until September. A typical locality is the Mbisi forest near Sumbawanga where it grows by streams in damp ground on the edge of the forest.

Pavonia urens Cav. [6B]

An erect herb or shrub 1.5 to 3 m tall, usually densely golden-brown pubescent on its stems and leaves with hairs that may occasionally cause a slight skin irritation. Leaves very variable in size, the lowermost up to 15 cm long and 18 cm broad but the upper much smaller, borne on stalks 2.5 to 11 cm long. The flowers, 3 to 5 cm in diameter, are bright rose-pink to mauve with a darker base and veins and have a central pink column of stamens 1.5 to 2.5 cm long.

This is a very variable species particularly in its leaf-shape, flower colour and pubescence as might be expected in a species found throughout tropical and South Africa and also in Madagascar and Réunion. In our area, it is a common plant of forest margins, secondary forest and scrub from about 600 to 2 000 m. It can be seen in flower from April until July, for example, in the Mufindi area, in the forests on Kito Mt near Sumbawanga and at Nyumbanyito a few kilometres north of Njombe.

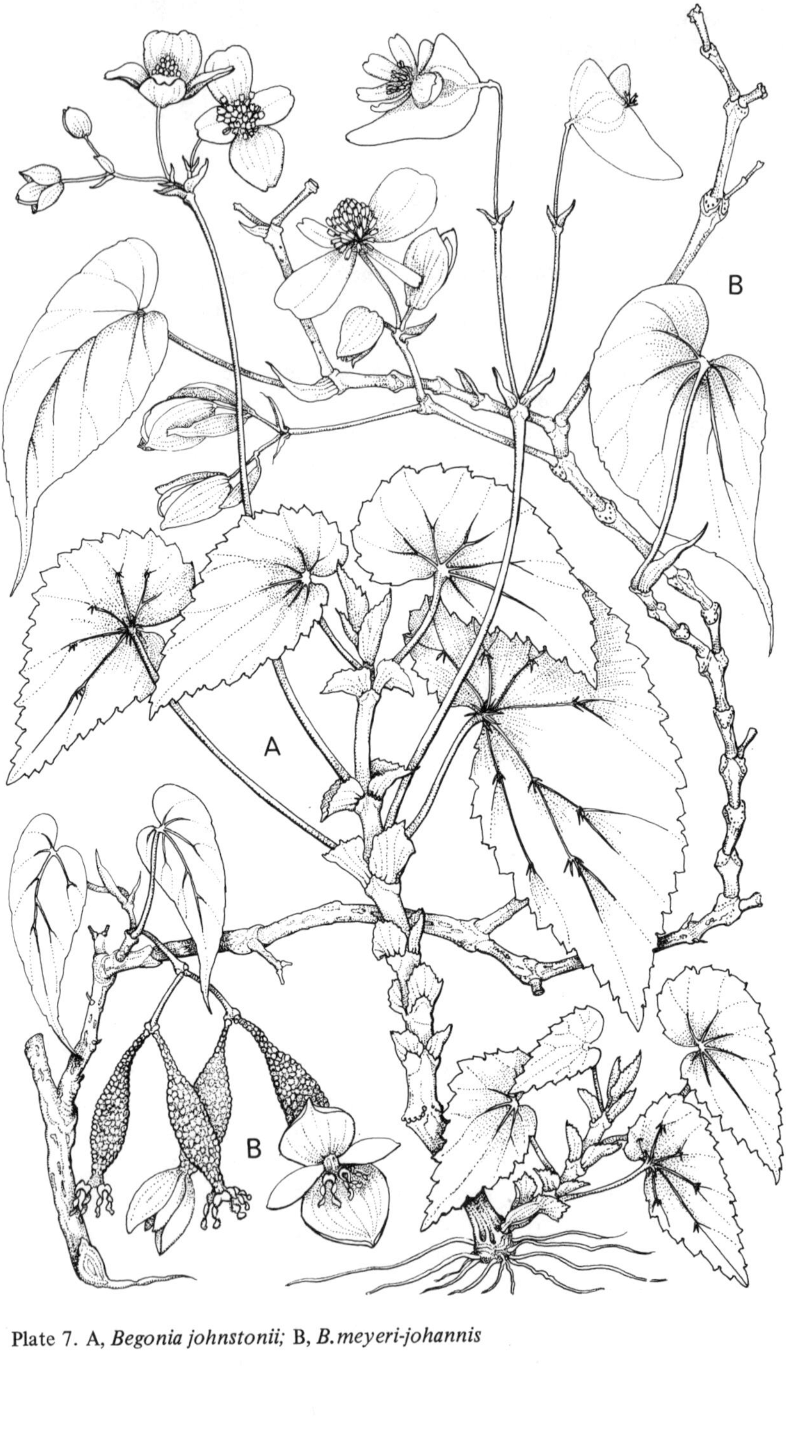

Plate 7. A, *Begonia johnstonii;* B, *B.meyeri-johannis*

BEGONIACEAE
THE BEGONIA FAMILY

Plates 7 and 8

Herbs commonly with succulent stems and large membranous stipules. Leaves alternate and asymmetrical, often attractively marked on their upper surface. The inflorescence bears separate male and female flowers which often differ markedly. In the male flowers, there are usually four petal-like segments around many stamens whereas the female flowers have an inferior trilocular ovary surmounted by more or less papillose and twisted stigmas.

Many *Begonia* species make attractive garden or pot plants.

Begonia johnstonii Oliv. ex Hook.f. [7A]

This attractive species produces thick, succulent, erect stems, 30 to 40 cm tall, which are commonly pale green or pinkish green and streaked with red. The leaves may reach 12.5 cm long and 5 cm broad. Flowers, usually 2.5 cm in diameter, white, pale pink or pink with bright yellow stamens and fruit red-winged.

It is a widespread species in East Africa from Kenya southwards and is common in the Usambara and Uluguru Mts of northeastern Tanzania. In the Southern Highlands, it is confined to the eastern side where it has been collected, for example, in the Gologolo Mts near Iringa growing at about 1 500 m on rocks below montane forest patches and in wet grass by streams. It usually flowers there from January until March.

Begonia meyeri-johannis Engl. [7B]

A large climbing species often reaching 7 m or more long. Leaves, 11 cm long and 6 cm broad, dark green above, red below with red stalks. Flowers about 3 to 4 cm in diameter, borne on red stalks, commonly pink or white with yellow, greenish yellow or white stamens.

A common plant in the montane forests of the Southern Highlands from Dabaga and Mufindi across to the Poroto Mts and Rungwe Mt, it can be best seen in flower during the dry season months of September and October. It grows commonly up forest trees by the track leading to the Kigogo River in Mufindi at about 1 900 m. By the track up Ngosi Mt in the Porotos it can be seen in the forest growing up trees of *Bersama, Ilex* and *Pygeum.* In Undali where it is also common it is called 'bosyo' by the Wandali people.

Elsewhere, it is widespread as far north as Uganda and Kenya and is also found in central Africa over an altitudinal range of 1 500-3 000 m.

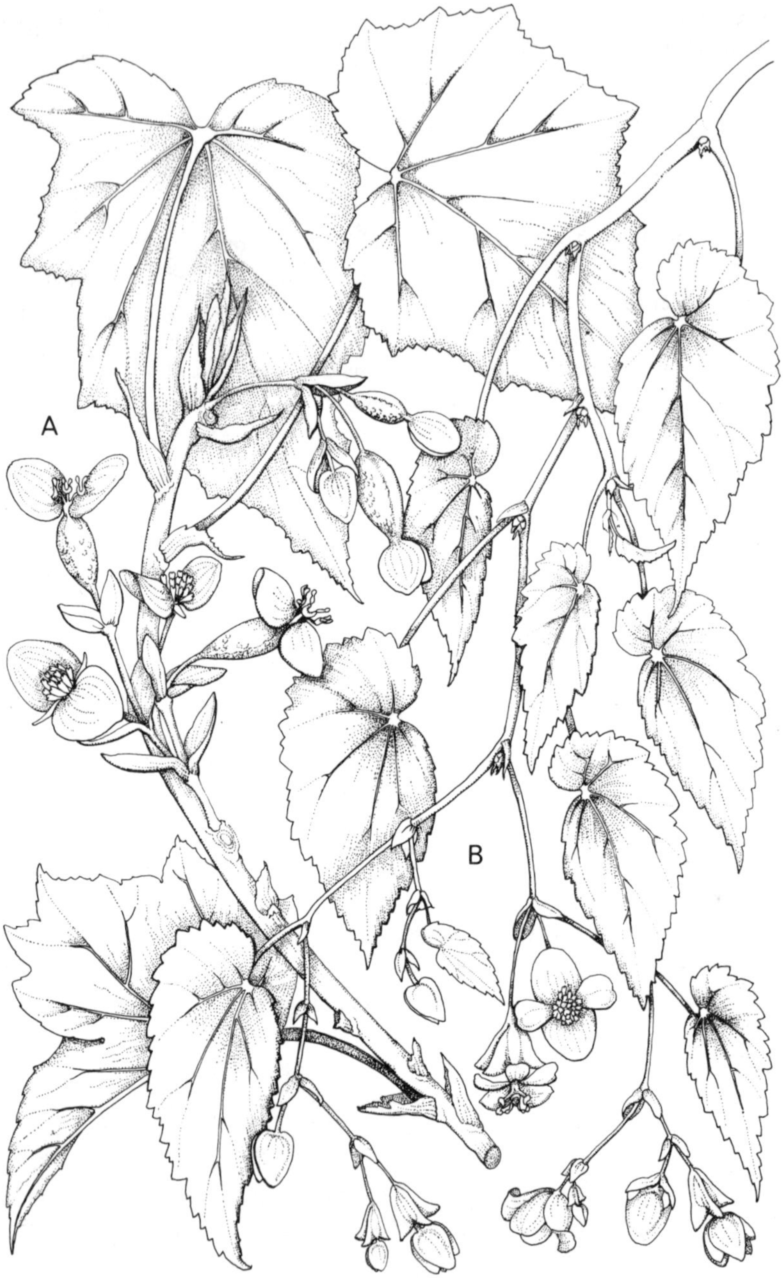

Plate 8. A, *Begonia oxyloba;* B, *B.sutherlandii*

Begonia oxyloba Welw. ex Hook.f. [8A]

Often a large plant with pinkish or dull red stems, flecked with white, up to 1.2 m long but usually rather shorter. Leaves 10 to 20 cm or more across, green with pink veins. Flowers, usually about 2.2 cm in diameter, are pink or white with pink venation and have yellow stamens.

Like *B.johnstonii,* it should be looked for on the eastern flanks of the Southern Highlands up to about 1 700 m altitude. It commonly grows on the banks of streams or on rocks by waterfalls in montane or lower montane forest.

A widespread species in Africa from the Cameroun, Uganda and Kenya south to Malawi and Angola.

Begonia sutherlandii Hook.f. [8B]

A pretty species with fleshy pink stems from 10 to 80 cm long often found hanging over rocks in damp situations. Leaves pubescent on the undersurface, commonly dark green and veined with red or, rarely, entirely reddish. Flowers, 2 to 2.6 cm in diameter, usually orange or orange-red in colour with yellow anthers.

It is common throughout the Southern Highlands and on the Mahenge Plateau in suitable localities. On the latter, we found it at the top of Ngongo Mt behind Sali growing under a rock overhang with the rare pink balsam *Impatiens joachimii.* Another locality where we have seen it is near Ibaba in the Bundali Hills where it cascaded in sheets over the rocks. It is also found on the magnificent Kiejo volcano which rises to the east of Tukuyu and is accessible where the road from Masebe to Manow, which is a large Lutheran mission station, crosses lava flows only some 200 years old. *B.sutherlandii* grows in the rock crevices of the lava along with *Streptocarpus solenanthus, Pentas purpurea* and *Cynorkis* species.

On the Kitulo Plateau, plants resembling this species but larger in all their parts, can be found on rocks above the stream which flows off the plateau on the east side of Ishinga Mt. Possibly this warrants recognition as a variety or subspecies.

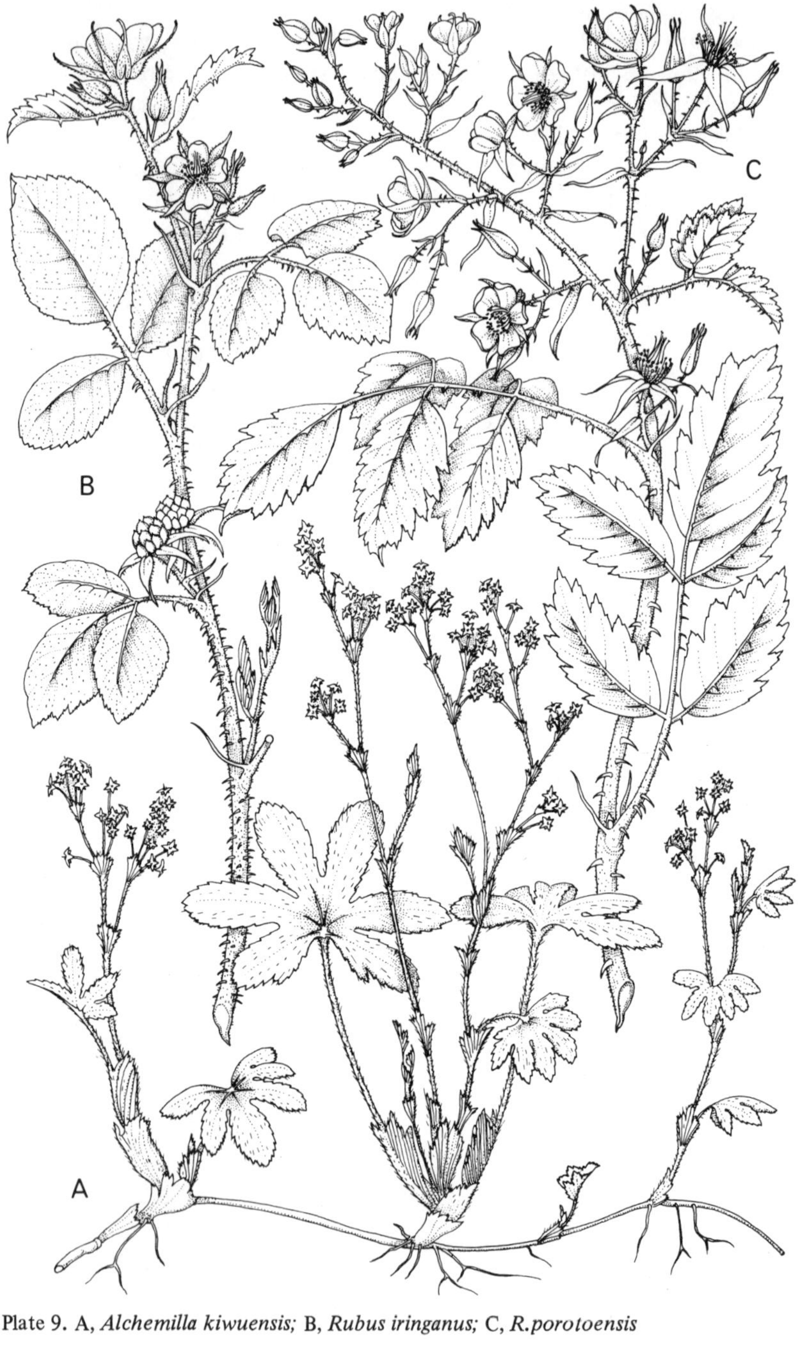

Plate 9. A, *Alchemilla kiwuensis;* B, *Rubus iringanus;* C, *R.porotoensis*

ROSACEAE
THE ROSE FAMILY

A large family of herbs, shrubs or trees; stems sometimes thorny as in *Rosa* and *Rubus.* Leaves alternate, simple or compound and typically with a pair of stipules at the base of the leaf-stalk. Flowers often large and showy, regular and bisexual and characteristically having an extra calyx below the regular one. Sepals and petals usually five and stamens and carpels numerous. Fruits diverse often fleshy and edible as in apples, plums, strawberries, blackberries and raspberries, common fruits in many parts of the world. Many species and hybrids of the rose, *Rosa,* are grown in gardens for their ornamental value.

The African species of Rosaceae are commonly mountain plants and are often species of genera far commoner in the temperate areas of Europe and Asia.

Alchemilla kiwuensis Engl. [9A]

Only three species of lady's mantle are found in the Southern Highlands of Tanzania, most of the other 14 East African species only reaching as far south as Kenya and northern Tanzania.

All the species in our area are creeping herbs with the characteristically lobed, soft, green leaves of the genus.

A.kiwuensis Engl., illustrated here, is a common and widespread species above 1 250 m. It has tiny flowers as in *A.cryptantha* but differs in having persistent tufts of basal leaves, 5- to 7-lobed stolon leaves and few-flowered inflorescences.

A typical situation where *A.kiwuensis* may be found is in the grassland by the margins of upland forest at Kiperu, Mufindi at 1 900 m and again in a similar habitat on Mt Rungwe at 1 700 m.

A.cryptantha has long slender stolons which root at the nodes and are covered with long whitish or yellowish hairs. The kidney-shaped leaves are five-lobed, densely hairy on both surfaces and range from 1 to 3.5 cm long and 1.7 to 4.5 cm broad. The tiny greenish yellow flowers, less than 2 mm across, have a hairy calyx and are more or less hidden by the stipules.

It is widespread in the Highlands growing in grassland or in moist bamboo thickets above 1 300 m. Elsewhere, it is widely distributed from Eritrea and the Sudan south to the Transvaal, also in Cameroun and in Madagascar.

A.ellenbeckii Engl., the third species, occurs in our area as subsp. *nyikensis* (De Wild.) R.Graham. Like *A.cryptantha* it lacks persistent tufts of basal leaves but has 3- to 5-lobed stolon leaves and a glabrous calyx tube. It grows in grassland on the margins of montane forest and in bogs and marshes from 1 470 to 1 800 m in the Southern Highlands and neighbouring Malawi and Zambia.

Rubus iringanus C.E.Gust. [9B]

About 18 species of bramble have been recorded from East Africa including at least two introduced species, namely *R.niveus* Thunb. from Asia readily recognised by its leaves which are whitish-grey felted beneath and its small purple flowers and *R.ellipticus,* another Asiatic species, now spreading rapidly in the Mbeya area.

Several other species are widespread in Africa but a few are rather restricted in distribution. *R.iringanus* is one of these being confined to the southfacing escarpment forests of the Mufindi district.

It is a dwarf bramble with slender stems covered with hooks and ochre-coloured spreading hairs and trifoliate leaves, each leaflet being up to 4 cm long and softly hairy on both surfaces. The flowers are white, up to 1 cm across, and its berries red.

It was first discovered in the Kigogo forest at Mufindi in January 1934 at 1 800-1 900 m altitude.

Rubus porotoensis R.Graham [9C]

Another bramble with a rather restricted distribution, it is a scrambling shrub with softly hairy prickly stems. The leaves are trifoliate or simple; the leaflets 4 to 8 cm long, rather sparsely hairy above but more hairy below. The prominent veins on the underside of the leaflets are characteristic. It has an almost pyramidal inflorescence which can reach 30 cm long and 20 cm across and bears many white to rose-pink flowers about 2.5 cm wide and, later on, red berries.

R.porotoensis is, as its name suggests, found on the edges and in clearings in the montane forests of the Poroto Mts and adjacent Mt Rungwe between 2 100 and 2 340 m. A typical locality is the forest on the slopes of the Ngozi crater above the village of Isongole.

In Usafwa *Rubus* species are called ‘udono’, in Ukinga ‘utomu’, in Umalila ‘itononkwa’ and ‘mtonongwa’, in Undali ‘lidongonelo’, in Uhehe ‘linunwa’ and in Ubena ‘litanula’.

VIOLACEAE
THE VIOLET OR PANSY FAMILY

Plate 10A

Herbs, shrubs or trees with alternate, entire or dentate leaves and small stipules. Flowers commonly regular (but bilaterally symmetrical in *Viola*) and bisexual. They have five sepals, petals and stamens, the last forming a ring around the superior ovary. In *Viola,* a nectar-containing spur is produced at the base of the lowermost petal. Fruit a capsule dehiscing into three or five valves.

Some species of *Viola* make attractive garden plants.

Viola abyssinica Steud. [10A]

A creeping herb with long trailing glabrous or hairy slender stems, up to 50 cm long, rooting at the nodes. The leaves, 1 to 4 cm long, are sparsely hairy below and borne on 1.2 to 2 cm long stalks. The flowers, borne on stalks 2 to 6 cm long at each node, are pale blue, lilac or violet with darker venation at the base of the 7 to 10 mm long petals, and have orange-yellow anthers and a greenish-brown spur, 3 to 4 mm long.

It is frequently seen in the mountains throughout the area from about 1 800 to 2 600 m altitude. In the Poroto Mts it is common in the patches of bamboo forest on Ngozi Mt where it flowers from October onwards. In Mufindi we have seen it in flower in January in the Kigogo forest while, in Ufipa, Mary Richards collected it growing on the bank of a stream at 2 500 m in the Mbizi forest in November. At Matamba, near Njombe, it grows in grassland at 1 800 m, flowering in January. It is widespread in Africa from Ethiopia southwards to South Africa and is found up to 3 000 m altitude on Mt Kilimanjaro in northern Tanzania.

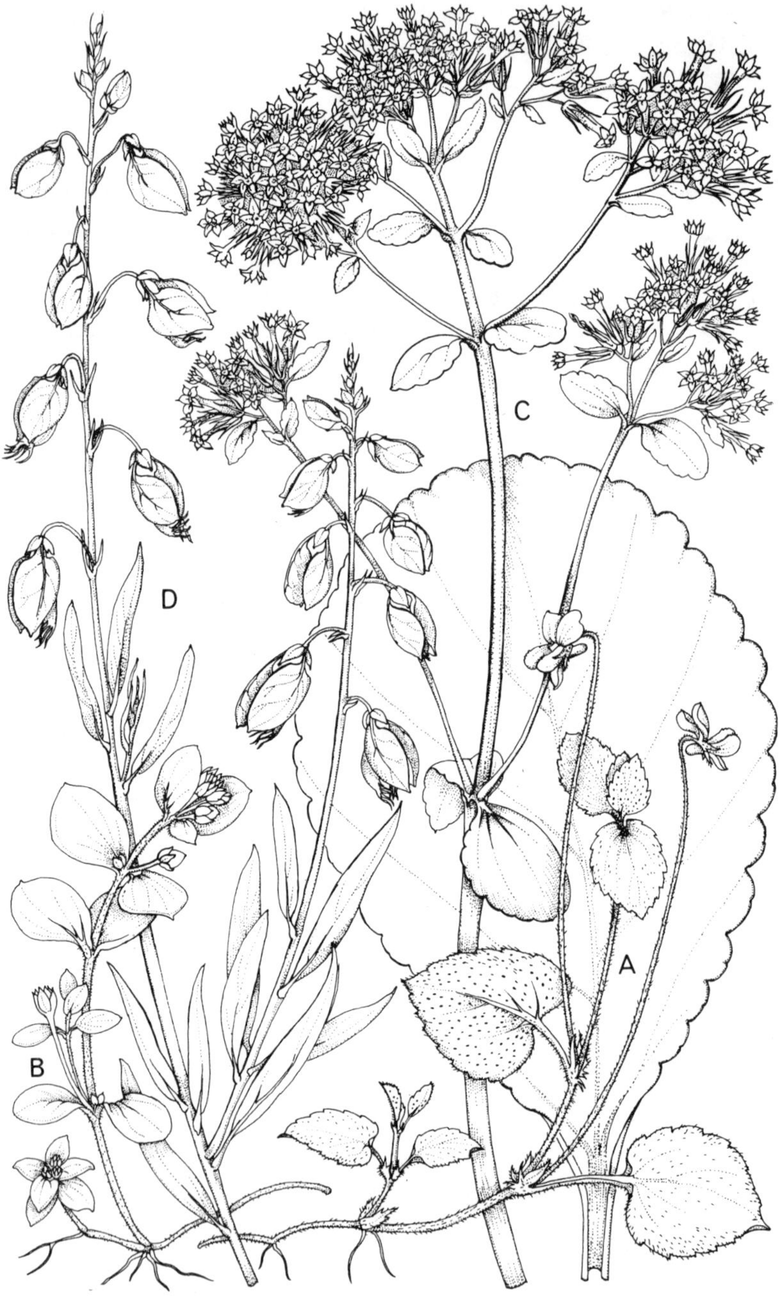

Plate 10. A, *Viola abyssinica;* B, *Crassula alsinoides;* C, *Kalanchoe densiflora;* D, *Polygala virgata*

CRASSULACEAE
THE STONECROP FAMILY

Plate 10B and C

Mostly perennial succulent xerophytic shrublets or herbs with fleshy entire leaves often borne in a rosette. Flowers structurally simple, generally with five free sepals, petals and carpels and one or two whorls of stamens; ovary superior.

The genus *Kalanchoe* contains several attractive species worthy of cultivation especially as pot plants.

Crassula alsinoides (Hook.f.) Engl. [10B]

This species forms mats of creeping stems, 12 to 25 cm long, which cover considerable areas of ground. Leaves shortly stalked, about 1.5 to 3 cm long and 1 cm broad. Inflorescence about 2 cm across with white flowers, 6 to 9 mm in diameter.

In the Rungwe forests, we have found it flowering in August and it has also been collected in the Upangwa Highlands at Milo and in Mufindi. It seems to prefer damp areas such as mud patches near streams and occurs over a wide altitudinal range from 1 500 to 2 650 m.

C.alsinoides is a relatively local species in the Southern Highlands and is also found in the northern mountains of Tanzania. The commonest species in our area is *C.alba,* an erect perennial, growing from a tuber, which produces a magnificent display of white, pink or red flowers during the wet season. Great masses of it can be seen on the Kitulo Plateau where it slopes down to the Unyakyusa forests.

Kalanchoe densiflora Rolfe [10C]

A succulent plant with a fresh green stem, up to 1.5 m tall, and bright green fleshy leaves, the lower ones up to 12 cm long and 10 cm broad but the upper ones smaller; leaf-stalks up to 4 cm long. Its inflorescence, up to 20 cm long, bears many tubular bright red or yellow flowers each about 1.2 cm long.

This is one of the species which brighten up the otherwise drab brown landscape of the dry season between August and November. Widespread in the mountains of the area above 1 800 m, it is particularly common on roadsides and on forest margins. The Wamalila call it 'itivwa'; the Wandali 'itifya'; the Wakinga 'm̄bokaboka'; and the Wanyakyusa 'inyongabembe'. All apparently use it medicinally.

A rarer *Kalanchoe* in the Southern Highlands, the yellow-flowered *K.lanceolata,* is found in the rocky terrain of the Kimani valley and Wangama escarpment above Kidugala flowering there in June and July.

POLYGALACEAE
THE MILKWORT FAMILY

Plate 10D

A family of herbs, shrubs, small trees and climbers with alternate usually simple estipulate leaves. Flowers bisexual, irregular, rather resembling those of peas and beans (*Leguminosae*). Calyx five-lobed, variously modified and often petaloid. Corolla usually reduced to three petals with the lowermost often saucer-shaped, sometimes with a fringed crest. Filaments eight, usually united at the base; ovary superior. Fruit commonly a capsule.

Polygala virgata Thunb. [10D]

P.virgata, the largest and most attractive species of the genus found in the Southern Highlands, is a branching shrub, commonly reaching 1.5 m tall but sometimes larger, with slender stems and glabrous or hairy branches. The bluish-green linear-lanceolate leaves are up to about 5 cm long and 6 to 15 mm broad. The flowers, borne in a terminal inflorescence 10 to 20 cm long, are a striking bright purple in colour with a dark purple tuft at the top. The floral segments are about 1 to 1.3 cm long.

In our area, it flowers from March until November and can be seen in the Mbeya Range on the Chunya road at about 2 600 m, and again in the Poroto Mts by the track to Ngosi Mt at 2 100 m. Further north, it grows near Chapota, Ufipa in the Empeta swamp at 1 650 m.

P.virgata is widespread in Africa as far south as South Africa.

LEGUMINOSAE
THE PEA OR BEAN FAMILY

Plates 11, 12 and 13

One of the largest families. Herbs, shrubs and trees usually with alternate, usually pinnate and stipulate leaves. The number of leaflets can be diagnostic, for example, all species of clover, *Trifolium,* have three leaflets. Most species have nodules on their roots which contain nitrogen-fixing bacteria. Flowers may be regular (in *Mimosa* and its relatives) but are more commonly bilaterally symmetrical. In the latter, the flowers have an upper petal called the standard, two lateral petals called the wings and two lower petals more or less fused into a keel and also commonly have 10 stamens. Almost all have a single carpel which is superior and develops into a pod.

Woody species of the family form an important element in the tree flora of Tanzania with *Acacia, Brachystegia* and *Albizzia* being particularly numerous in the open woodlands at lower and intermediate altitudes. Shrubby and herbaceous species are equally numerous in the flora and a small selection of these have been included here.

Many of the herbaceous species are of economic importance as food and fodder plants. Notable among these are the many different edible beans and also the pea. Their nitrogen-fixing properties also make them useful particularly in nitrogen-poor soils which are enriched by their presence.

Adenocarpus mannii (Hook.f.) Hook.f. [11A]

The only species of the predominantly Mediterranean genus *Adenocarpus* found in Africa is *A.mannii,* a plant widespread in the continent from Ethiopia and the Sudan across to Nigeria and southwards to Malawi, Zambia and Angola.

It is an erect shrub up to 4.5 m tall with densely hairy much branching stems. The trifoliate leaves are clustered on short stalks, the leaflets being hairy below and up to about 8 mm long. The inflorescences consist of a cluster of 6- to 20-flowers in a head. The flowers are yellow and about 8 to 9 mm long, emerging from a hairy calyx. The pods are hairy and about 1.8 to 2.5 cm long.

A.mannii is widespread in the Southern Highlands area growing from 1 500 m upwards on the margins of montane forests, in upland grassland and moorland and even on rocky mountain tops. A typical locality where it can easily be seen in flower from June until September is on Mbeya Mt.

Agyrolobium stolzii Harms [11B]

A perennial herb with slender hairy stems 25-75 cm tall. Leaflets 2.5-4.5 cm

Plate 11. A, *Adenocarpus mannii;* B, *Agyrolobium stolzii;* C, *Crotalaria recta;* D, *Dolichos sericeus;* E, *Cassia mimosoides*

long, hairy on both surfaces. Flowers bright yellow, about 1 cm long. Pods 2-4.5 cm long, densely covered with appressed hairs.

A common species of the Mbeya and Rungwe regions growing in grassland often by rivers and marshes, from 1 350 to 2 200 m.

Cassia mimosoides L. [11E]

Nearly 50 species of the attractive legume genus *Cassia* grow in East Africa and 20 have been recorded in the Southern Highlands area. Many are native species but some such as the shrubby *C.floribunda* Cav. are undoubtedly introduced. *C.mimosoides,* illustrated here, is one of the most widespread species in the genus, found throughout the tropical regions of the Old World and possibly also in tropical America. It is an exceedingly variable prostrate to erect herb commonly with hairy stems up to 1.5 m long and leaves 0.6 to 10 cm long and 0.4 to 1.5 cm broad, parallel sided, with from 16 to 76 pairs of ciliate leaflets. The inflorescences emerge either from or above the leaf axils, each bearing 1 to 3 yellow flowers with petals 4 to 13 mm long. The pods are usually hairy and 3.5 to 8 cm long.

Several variants of *C.mimosoides* have been recorded from the Southern Highlands area but these have not been given formal taxonomic recognition. One of the earliest collections was made by Stolz at Kyimbila in September 1910. We have seen it in flower in many places throughout the area at all altitudes up to about 1 600 m.

One of the most prominent species of *Cassia* in the area is the introduced and widely planted tree *C.siamea* Lam. This flowers in April and can be seen to great effect along the main road between Mbeya and Iringa.

Crotalaria recta A.Rich. [11C]

One of the largest of all genera in the East African flora is *Crotalaria* with almost 200 species recorded so far. It is characterised by its inflated pods, its flowers in which the keel petals are beaked and the five basally attached anthers alternate with five shorter dorsally attached ones. Needless to say, with so many species present naming is not easy other than to genus.

A common species in our area is *C.recta* a sparingly branched perennial, 1 to 2.7 m tall, with hollow, ribbed, usually hairy stems and trifoliate leaves which are more or less hairy beneath. The leaflets are 1.8 to 14 cm long and 2.5 to 6 cm broad. The inflorescences, 20 to 50 cm long, bear many, rather clustered, yellow flowers which are veined with purple or maroon at the base within and on the outer surface. The lower petals (keels) are 1.2 to 1.5 cm long and are slightly shorter than the wing petals. The pods are glabrous, 5 to 6 cm long.

In our area *C.recta* is commonly found growing in grassland and on forest margins although elsewhere, at lower altitudes down to 850 m, it prefers marshes and is also found as a weed of cultivation. A typical locality for *C. recta* is the grassland by the road leading to Mufindi from the main Mbeya to Iringa road. It flowers from March until about July.

Plate 12. A, *Indigofera atriceps;* B, *Kotschya recurvifolia;* C, *Lotus goetzei;* D, *Trifolium simense*

Elsewhere in Tanzania, *C.recta* is widespread and its distribution extends north to Eritrea, west to Nigeria and south to the Transvaal.

Dolichos sericeus E.Mey. [11D]

To the untrained eye few flowers can be identified from a car or motor-bike travelling at any speed along the mountain roads. However, the bright flowers of *D.sericeus* make it an exception for it stands out from the surrounding vegetation on roadsides throughout the area.

Over 20 species of *Dolichos* have been found in East Africa and nine of these are known from the Southern Highlands area. Of these, *D.sericeus* is possibly the most variable and also the commonest species. It is a climbing or prostrate herb with slender twining hairy stems 60 cm to over 2 m long. The hairy leaves are trifoliate, each leaflet being from 1 to 8.5 cm long. The 1- to 10-flowered inflorescences in the leaf axils have hairy stalks up to 18 cm long. The flowers are pink, magenta or bluish-purple and are mostly up to 1.8 cm long and wide.

Four subspecies occur in East Africa but only the typical one is found in our area, distinguished by its glabrous pods, hairy calyx and medium-sized flowers. We have seen it in mid-June growing by the road in grassland at 2 300 m near Idunda in the Umalila Hills. It is also common in the Poroto Mts, in the Mbeya Range on Loleza Mt just behind Mbeya and on Rungwe Mt flowering from June until August. Elsewhere, it is widespread from the Sudan and Ethiopia southwards to Natal and across to Cameroun in the west.

D.ichthyophone Verdc., a purple-flowered species with simple leaves and trifid bracts is restricted to Ufipa where it was collected several times in the grasslands between 1 500 and 2 400 m by Mrs Richards. *D.cardiophyllus* Harms, another simple-leaved species with predominantly purple flowers, is restricted to grasslands near Iringa.

Indigofera atriceps Hook.f. [12A]

If you see a small red-flowered member of the pea family in the mountains of the Southern Highlands it is very likely to be one of the 60 or so species of *Indigofera* recorded from the area. One of the species most frequently seen is *I.atriceps.* It is a coarse, branching, usually erect herb up to 2 m. The branches, leaf stalks and seed pods are glandular hairy and the flowers are red or pink. This species is extraordinarily variable with seven subspecies in East Africa. Three of these occur in the Southern Highlands being found mostly between 1 000 and 3 200 m altitude in a variety of habitats from upland grassland and montane forest margins (subsp. *atriceps*) to swampy grassland and old cultivation areas (subsp. *setosissima* (Harms) Gillett) and to *Brachystegia* (miombo) woodland (subsp. *glandulosissima* (R.E.Fries) Gillett). The plant figured is subsp. *atriceps* which has long hairs on the pod and longer filaments and pods than subsp. *setosissima.* Subsp. *glandulosissima* has a pod densely covered with short glandular hairs.

Considered as a whole this species is widespread in tropical Africa south to Malawi, Zambia and Mozambique.

Of the other species which brighten the mountain slopes of our area, *I.mimosoides* is particularly common whilst *I.smutsii, I.ramosissima, I.asterocalycina* and *I.spathulata* are all endemic to the region. If you see a bright carpet of red on any roadside in the mountains, stop and you will probably find there a dense mass of *Indigofera* in full flower. The Wamalila call these plants 'ivizi' and the Wanyiha, 'vumamkola' and 'ushiboku'.

Kotschya recurvifolia (Taub.) F.White [12B]

About 30 species of *Kotschya* are found in tropical Africa and Madagascar and 14 of these occur in the Southern Highlands. In some areas of the Highlands the species illustrated here, *K.recurvifolia,* is one of the dominant plants, particularly in the upland grassland above 1 800 m of Umalila, Rungwe and Ufipa and between Iringa and Njombe. Bees love it when it is in full flower in July. The Wamalila call it 'intenga', the Wabena and the Wakinga call it 'matenga', the Wahehe, 'lwelesi' and the Wandali 'ndenga'. It is a very variable, erect, much branched, somewhat aromatic and sticky subshrub up to 45 cm tall. The stem is pubescent and covered with long whitish or golden bristly hairs. The densely flowered inflorescences are also covered with golden bristly hairs, the flowers being golden yellow with a greenish keel. The pods are hairy and the seeds greenish to dark red-brown.

Three subspecies, two of which occur in our area, are known. Subsp. *recurvifolia,* with congested foliage and small flowers, is common in upland grassland and on forest margins usually on volcanic soils between 1 800 and 3 000 m. Subsp. *longifolia* Verdc. has less congested foliage so that the stems are visible and larger flowers and it is found in the open parts of montane forest about 1 800 m.

The species as a whole is widespread from Ethiopia south to Malawi and Zambia.

Other *Kotschya* species in the Southern Highlands include *K.goetzei,* endemic to the central Mbeya Mountains, *K.platyphylla,* restricted to the eastern Highlands, and two blue-flowered species, *K.strigosa* and *K.caelescens.*

Lotus goetzei Harms [12C]

This species was named after Goetze who was one of the first botanical collectors to penetrate the Southern Highlands. He first collected it in the Poroto Mts just southeast of Mbeya and he died shortly afterwards near Tukuyu in 1899. It is extremely variable in all its features, ranging from a prostrate plant to a small bush over 1 m high. It is also very variable in its leaflet shape and in its pubescence. Its inflorescences are 1- to 6-flowered umbels and each flower has a red calyx with white petals marked with pink or purple-red.

August is not normally the best time to look for flowers in the Southern

Highlands for it is the middle of the dry season. *Lotus goetzei* is, however, at its best then and brightens up an otherwise drab countryside. One area where it is particularly common and conspicuous is the Ishipi valley near Isonso on the edge of the Undali Hills where they break away from the Umalila Plateau. Its pretty red and white flowers stand out on the hillsides where the plant forms dense stands in association with various species of Compositae such as *Vernonia syringifolia* and also the bushy *Gnidia involucrata* and the peas *Eriosema ellipticum* and *Tephrosia aequilata. L.goetzei* is called 'idava' by the Wamalila. It is a true mountain flower, rarely seen below 2 000 m.

L.goetzei is widespread elsewhere in East Africa from Ethiopia south to northern Malawi and also possibly across to Angola.

Trifolium simense Fresen. [12D]

Some 22 species of clover (*Trifolium*) are found in East Africa and about nine of these have been recorded from the Southern Highlands. All can be recognised by their trifoliate leaves and somewhat globular or ovoid flower-heads.

Trifolium simense is one of the most widespread being found in upland grasslands above 1 500 m from Cameroun across to Ethiopia and Sudan in the north and southwards to northern Zambia and Malawi. A low creeping plant it is readily recognised by its narrowly linear leaflets with toothed margins. Its flower heads are about 1.5 cm across with glabrous or sparsely hairy white bracts and reddish purple (or rarely white) flowers.

We have seen *T.simense* in flower at the end of March in the upland grasslands of the Umalila Plateau near Ilembo. The Plateau there is at about 2 250 m altitude and is dissected by many rivers in valleys about 150 m below the plateau surface. Along the sides of these valleys a great variety of flowers, including *T.simense,* can be found growing in the rich volcanic soils. The richness of these soils has led to large areas being ploughed in recent years for pyrethrum, millet and maize cultivation.

Elsewhere in our area, *T.simense* is common in the Mbeya Range and on the Nyika Plateau in neighbouring Malawi.

Other clovers found in the region are the Ukinga endemic *T.ukingense* and *T.wentzelianum,* another species of restricted distribution known only from the central highlands around Mbeya and Njombe.

Lablab purpureus (L.) Sweet [13A]

The delightfully named *Lablab purpureus,* a widespread plant in East Africa from sea level up to 2 400 m, is a climbing perennial herb with 1 to 5 m long stems with glabrous or hairy trifoliolate leaves and leaflets from 2.5 to 15 cm long. The few- to many-flowered stiff elongate inflorescences are borne in the leaf axils. The flowers are usually reddish purple although pale purple- or white-flowered forms are not infrequent with petals 1.2 to 1.8 cm long and 10

Plate 13. A, *Lablab purpureus;* B, *Tephrosia interrupta;* C, *Vigna pygmaea*

stamens, nine of which are joined while one is free. The pods are flattened and about 4 cm long.

Three subspecies of *L.purpureus* have been recognised. Subsp. *purpureus* and subsp. *bengalensis* (Jacq.) Verdc. are both cultivated forms with large pods. Both have been cultivated occasionally in East Africa but neither is an important crop plant here. The Tanzanian plants are all referable to subsp. *uncinatus* Verdc. a small-podded truly wild plant which is common at all altitudes up to 2 400 m in our area.

Tephrosia interrupta Engl. [13B]

A variable branched shrub, 3 to 5 m tall, with brown-pubescent young stems and imperfectly pinnate leaves up to 11 cm long, the leaflets pubescent on their lower surface. Flowers purple, 1.4 to 2.1 cm long, in terminal tomentose pseudo-racemes or subumbels. Pod densely grey- or brown-tomentose, up to 9 cm long and 5 to 6 mm wide.

It is widespread in Africa from the Sudan and Ethiopia south to Malawi, three subspecies being recognised, all of which are found in the Southern Highlands; the typical one up to 3 m tall with a terminal inflorescence of several nodes, small flowers 11 to 16 mm long and pods up to 8 cm long; subsp. *mildbraedii* (Harms) Gillett, similarly small-flowered, reaching 5 m tall, with a reduced almost subumbellate inflorescence and pods up to 7 cm long; and subsp. *elongatiflora* Gillett with long inflorescences, large flowers 17 to 20 mm long, and pods up to 9 cm long.

All are found in upland grassland and on forest and scrub margins mostly above 1 500 m altitude and flower in August and September. The typical subspecies illustrated here has been found in mid-August near Isuto in Umalila at 1 750 m. The Wamalila call it 'bazivanga' and Wasafwa 'ibanga'.

About 30 species of *Tephrosia* have been recorded from the Southern Highlands and reference should be made to the *Flora of Tropical East Africa* for a key to their identification. *T.kalamboensis,* collected by Mary Richards in 1965 at the Kalambo Falls, has been found nowhere else while several others, namely *T.lepida, T.paucijuga, T.muenzneri* and *T.iringae* are all endemic to the area. In contrast *T.vogelii,* a branching shrub to 4 m tall with white flowers and formerly widely used as a fish-poison, is the most widespread but probably not native.

Vigna pygmaea R.E.Fries [13C]

Nearly 40 species of *Vigna* have been recorded from East Africa and nearly half of these have been found in the Southern Highlands area. A few species produce edible seeds and are cultivated notably *V.unguiculata* (L.) Walp., the cowpea.

V.pygmaea, one of the wild species found in the woodlands of the area from 1 500 to about 2 300 m, is a short erect herb with several stems, 3 to 30 cm tall, flowering when quite leafless. The hairy stems emerge from woody carrot-like rootstocks. When they appear the leaves are trifoliate and bristly hairy with leaflets 1 to 6.5 cm long. 1- to several-flowered inflorescences covered with brownish or blackish hairs are terminal or produced from the leaf axils and reach 18 cm long. The flowers are purple and 1.5 to 2.1 cm long, the calyx covered in brownish or black hairs. Its densely grey or brown hairy pods reach 9 cm long and contains several smooth blackish seeds.

This species is at its northern limit in the Southern Highlands. We have found it near Songwe in the Umalila Highlands flowering in August in burnt areas at 1 900 m. This area, in common with much of Africa, is burnt by the local people before the rains and this always triggers off the flowering of many plants. At Songwe it grows in association with the tiny red-flowered milkwort *Polygala ohlendorfiana, Euphorbia cyparissioides, E.depauperata, Aster tansaniensis* and *Gnidia kraussiana.*

Elsewhere *V.pygmaea* is widespread in south-central Africa and is also found in Cameroun. Var. *grandiflora* Verdc. distinguished by the spreading rather than appressed hairs on its stem and larger flowers with a standard petal nearly 2 cm long is found in Malawi and Zambia and also in Ufipa where Mary Richards found it on Kito Mt in flower in mid-September.

THYMELAEACEAE
THE DAPHNE FAMILY

Plate 14

Mostly shrubs with stringy bark, entire simple leaves and lacking stipules. The flowers are regular and usually bisexual with parts in four or fives and are basically cup-shaped with the floral parts arranged around the rim. The sepals are petaloid with the stamens inserted on the tube. The petals are inconspicuous. The ovary is superior and usually of one or two fused carpels.

The major African genus of the family is *Gnidia* with about 100 species. Species of the allied genus *Daphne* are popular in cultivation in temperate zones and in those areas with a Mediterranean climate.

Gnidia glauca (Fresen.) Gilg [14A]

A shrub or, less commonly, a small tree to 13 m tall with woody fibrous-barked stems, glabrous leaves 3 to 6.5 cm long and 0.8 to 1.6 cm broad, and inflorescences 3.5 to 5 cm in diameter. Flowers faintly sweetly scented, golden-yellow or orange-yellow, 1.2 cm long.

It is common throughout the Southern Highlands above about 2 000 m. In Mufindi, where it is called 'livana' by the Hehe people, it grows on the margins of the Kigogo Forest and also in the adjacent grasslands. In the Poroto Mts it is found in forest patches of *Hagenia, Podocarpus* and *Arundinaria* up to 2 650 m whilst around Njombe it is not uncommon in bushland growing with *Kotschya* and *Stoebe.* The Safwa people in the Mbeya Range call it 'malulu'. It can be found in flower from June until about October.

Elsewhere it is widespread throughout tropical Africa from the Sudan and Ethiopia south to Malawi and Zambia.

Gnidia involucrata Steud. ex A.Rich. [14B]

A erect perennial herb growing from a creeping woody rhizome. The glabrous, and commonly unbranched stems are up to 40 cm tall but may branch and reach 2 m in plants growing at lower altitudes. Leaves are 1.2 to 1.5 cm long and 1.5 to 4 mm broad. Yellow or creamy-yellow flowers, about 1.5 cm long, are subtended by large reddish or brown bracts, 8 to 10 mm long.

This species is widespread in tropical Africa from Ethiopia and West Africa south to Zimbabwe. In our area, it is common in the seasonally burnt grasslands above about 1 500 m flowering from about September until January. It can be seen in Ufipa near the top of the escarpment by the Mbizi Forest at 2 350 m, on the Chunya escarpment behind Mbeya at 2 250 m and at a lower altitude by the Mbozi circle road.

Plate 14. A, *Gnidia glauca;* B, *G.involucrata;* C, *G.kraussiana*

It is perhaps better known here under the later synonyms *G.buchananii* Gilg and *G.macrorrhiza* Gilg.

Gnidia kraussiana Meisn. [14C]
An extremely poisonous, perennial, silvery-pubescent herb with weakly branching stems arising from a woody rootstock. The variably pubescent leaves, 2 to 4 cm long and 0.5 to 1.2 cm broad, have translucent veins. Inflorescences, 3.5 to 5 cm in diameter, contain lemon-yellow to golden-yellow flowers 1 to 1.5 cm long.

In Mufindi, where the Hehe people call it 'linyanungwe', it grows in the seasonally burnt grassland areas such as along the road to Sawala and Ngwazi at 1 850 m. It can also be seen at Milo near Njombe at 2 300 m and in Ufipa near the Mbizi forest at 2 350 m growing in similar situations. It flowers between September and November.

Elsewhere it is widespread in tropical Africa from the Sudan south to Zimbabwe.

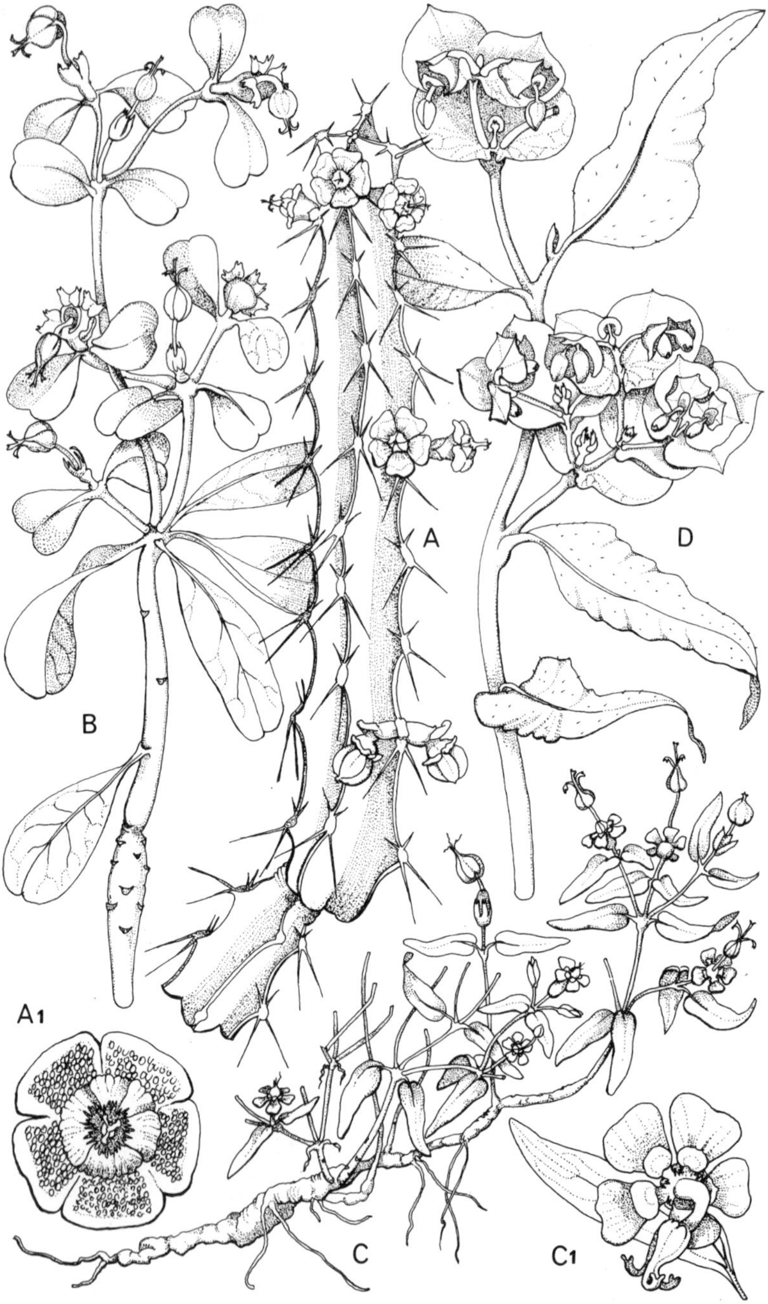

Plate 15. A, *Euphorbia quadrangularis* (A1, inflorescence); B, *E.transvaalensis;* C, *E.zambesiana* (C1, inflorescence); D, *Monadenium laeve*

Herbs or succulent shrubs and trees often producing copious amounts of milky latex which is sometimes poisonous. Leaves mostly simple and alternate with stipules which are often spiny. Flowers regular, unisexual and either occur on the same plant, as in *Euphorbia,* or on different plants and usually have a perianth of five segments. Stamens in male flowers one to many around a sterile pistillode. Ovary in female flowers superior, usually of three fused carpels. Fruit usually a schizocarp. Glands are a feature of the flowers of *Euphorbia* and these either encircle or are situated among the stamens or encircle the ovary. The inflorescence in *Euphorbia* is called a cyathium and consists of a small cup-shaped involucre bearing a number of glands around its rim with a number of male flowers surrounding a single central female one.

In Africa, many species of *Euphorbia* are xerophytic and resemble cacti. The best-known member of the family is *Hevea brasiliensis,* the source of the world's natural rubber.

Euphorbia quadrangularis Pax [15A]
A succulent plant to 3.5 m tall with sparsely branching 4-angled stems. Branches 1.6 to 2 cm broad, wide-spreading, often yellow-green marked with grey and liver-coloured patches and with angles bearing black-tipped spines. The relatively insignificant glossy, green flowerheads are about 7 mm in diameter.

It is a relatively low-altitude species in our area growing, from about 750 m upwards, in open bush or woodland, grassland or open stony soils. It is not infrequent in the Ruaha National Park and can also be seen at Mbarika near Iringa. It usually flowers in the dry season from July to August.

Euphorbia transvaalensis Schltr. [15B]
Copious amounts of white milky latex issue from the damaged leaves and stems of this succulent plant. It is rather variable in stature often forming a shrub up to 1.6 m tall with grey fleshy stems arising from a woody rootstock. Its grey- or yellow-green leaves, borne in whorls of 3 to 5, may reach 17 cm long and 6 cm broad. The flower-heads are rather insignificantly coloured pale or dull yellow-green and are 2 to 3 cm across.

Mary Richards collected the species in Ufipa several times, for example at Chala where it was growing in muddy patches in a dried-up stream-bed at 1 590 m and on the Mpui escarpment at 1 900 m. It can also be seen on the roadside between Mbeya and Chimala, some 23 km from the former, growing in rocky ground by a river. It flowers from December until March in our area. Elsewhere it is widespread from Kenya south to South Africa (as its name suggests).

Euphorbia zambesiana Benth. [15C]

This pretty spurge is a tiny plant with trailing red stems, varying from 4 to 15 cm, emerging from a woody rootstock. Its narrowly elliptic leaves, 8 to 16 mm long, are bright green above and reddish below. The flower-heads which peep out from the leaf-axils are 3 to 5 mm across. The female flowers are red and the male flowers white with a red disc.

It is commonly found in burnt ground growing in association with such plants as *Scutellaria paucifolia, Vernonia praemorsa, Ageratinastrum goetzeanum* and *Osteospermum monocephalum.* It is widespread in the upland grasslands around Iringa, Njombe, Mbeya and Sumbawanga. A typical locality is at the foot of Irundi Mt in Mufindi where it can be seen in flower in late October. It is also common on both the Kitulo Plateau and Mbeya Range from about 1 700 m upwards.

E.zambesiana is as its name suggests widely distributed in south-central Africa across to Angola and is also found in the mountains of Uganda.

Monadenium laeve Stapf [15C]

Several unbranched stems, 15 to 45 cm long, are produced by the woody rootstock of this unusual plant. It has yellow-green leaves up to 16 cm long and 3 to 4.5 cm broad which have short hairs on their upper surface but are glabrous below. Its flowers are pale green with a yellow crescentic disc at the base of the female flower and are enclosed by a wavy-edged bract-cup 6 to 14 mm in diameter.

Mary Richards found this species in March in Ufipa growing in crevices in the rocks of the Bukoban sandstone escarpment overlooking Kasanga on Lake Tanganyika. It has also been collected in May on the road from Mbeya to Lake Rukwa near Njerenje growing in *Brachystegia* woodland on a river bank at about 1 700 m altitude.

It is also known from neighbouring parts of Malawi.

Rather commoner in our area is the more widespread species *M.chevalieri* Denis which grows mostly above 1 700 m. It has distinctive narrowly linear glabrous leaves, 3 to 7 mm broad, and flowers subtended by bracts 4 to 5 mm long.

GERANIACEAE
THE GERANIUM FAMILY

Plates 16A and B

A large family of annual or perennial herbs or, more rarely, shrubs with opposite or alternate, simple or compound leaves often glandular hairy and stipulate. Flowers bisexual, regular or slightly irregular and either solitary or in cymes. Sepals and petals five, the latter often large and brightly coloured, surrounding two or three whorls of five stamens (and/or staminodes), usually with their filaments fused at the base. Ovary superior, of five fused carpels each with a separate style. Fruit a schizocarp, often elongate and dehiscing from the base upwards.

The genera *Geranium, Erodium* and *Pelargonium* are cultivated for their ornamental value. Several species of the latter genus are also grown for their aromatic oils widely used in perfumery.

Geranium arabicum Forsk. [16A]

Geranium is a common genus in Europe and temperate Asia but only seven species have been recorded from East Africa and all of these are mountain plants growing in the cooler climates found at higher altitudes.

G.arabicum is an extremely widespread and variable species found as its name suggests in Arabia and also in Africa from Egypt throughout the mountains of the tropics and in Madagascar. It is a branching trailing, ascending or rarely tufted perennial herb up to 90 cm long. Its creeping stems root at the nodes, are somewhat woody below and are longitudinally grooved above. It has hairy five-lobed leaves which are purple beneath and are borne on 4 to 26 cm long stalks emerging from between entire stipules. The inflorescence is generally 2-flowered and the 1.2 to 2 cm diameter flowers are white or lilac with red veins on the petals.

Two subspecies have been recognised. The first subsp., *latistipulatum* (A. Rich.) Kokwaro with leaves in which the lobes are divided to the base, is found in Ethiopia, Kenya and northern Tanzania. Subsp. *arabicum* with less completely divided leaves is found throughout the range of the species including our area. It grows in the Poroto Mts, on Rungwe and Mbeya Mts and on the Kitulo Plateau from 1 800 to 2 350 m. It flowers over a long period from August (in Mufindi) to January in the west and although it grows in a variety of habitats it is mostly readily seen in grassland or on moorland usually in moist areas near streams and in marshes.

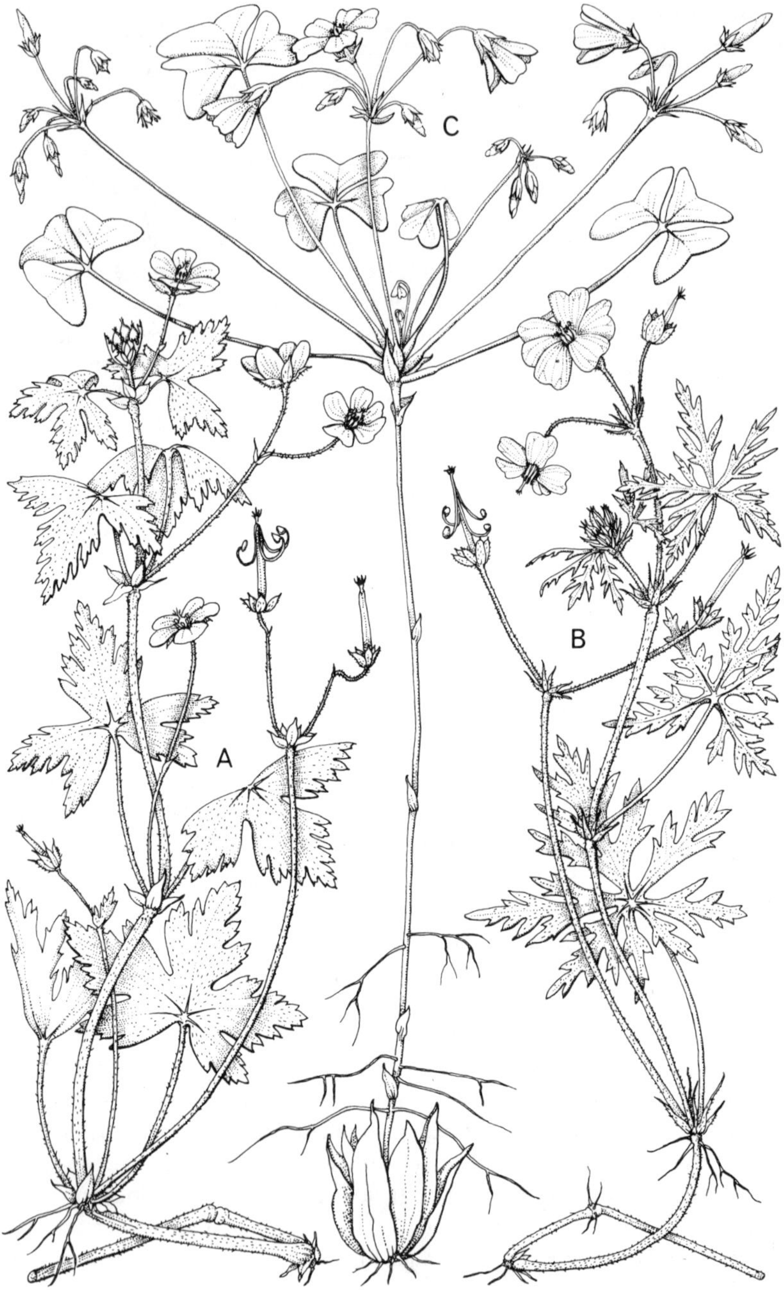

Plate 16. A, *Geranium arabicum;* B, *G.incanum;* C, *Oxalis semiloba*

Geranium incanum Burm.f. [16B]

G.incanum is another variable species common throughout the Southern Highlands. It is a semiscandent, creeping or rambling herb reaching 1.2 m long, with a hairy or glabrous reddish stem and 3- to 7-lobed deeply divided leaves, hairy above and silvery hairy below. The inflorescence is generally 2-flowered, the flowers being borne on glandular-hairy stalks 1 to 3 cm long. They are variable in colour having white, pink or mauve petals, 6 to 15 mm long, with darker veins and purple anthers.

As in the previous species, two subspecies have been recognised both occurring in the Southern Highlands. The typical one also recorded from South Africa, has linear leaf segments and red glandular hairs on the younger parts. It can be seen growing in montane grassland in the Dabaga Highlands at Kibengu, in the Mdapo marsh near Njombe and on the Kitulo Plateau. It flowers there from January until March.

Subsp. *nyassense* (Knuth) Laundon has oblong-lanceolate leaf segments and brown glandular hairs on the younger parts. It grows in upland grassland, along streams and in marshes from 900 to 2 600 m in the area, flowering from January until August. It can be seen in the Ukinga Mts near Mwakaleli, in the Mufindi region and elsewhere in suitable habitats. This subspecies is the only one recorded from south-central Africa.

OXALIDACEAE
THE BERMUDA BUTTERCUP FAMILY

Plate 16C

Annual or perennial herbs with alternate estipulate leaves which are commonly compound. In *Oxalis* the leaflets fold down in cold weather and at night. The flowers are regular, bisexual and have five free sepals and petals. There are ten stamens in two whorls around a superior ovary with five free styles. The fruit is a capsule.

The best known species of the family are wood sorrel, *Oxalis acetosella* and *Oxalis pes-caprae*, the Bermuda buttercup, a widespread, noxious, but very attractive weed of areas with a Mediterranean or subtropical climate.

Oxalis semiloba Sond. [16C]

A dozen species of *Oxalis* occur in East Africa, seven of these being recorded from our area. *O.semiloba* is one of the commonest. It is a perennial herb, 5 to 25 cm tall, growing from a 1 to 4 cm long bulb. The three-lobed leaves are borne on long stalks in a cluster at the apex of the stem. These are dark green above, each leaflet being 0.5 to 1 cm long and 1.2 to 2.5 mm broad. The sub-umbellate inflorescences reach 15 cm long and each has up to six flowers at the apex of the scape. The pendulous flowers are pale mauve to pink-crimson or bright mauve with a pale green style and stamens. The petals range from 10 to 15 mm long.

Two subspecies of this pretty *Oxalis* occur in the Southern Highlands. Subsp. *semiloba* has a sharply angled sinus whilst subsp. *uhehensis* has a wide rounded sinus to each leaf lobe.

O.semiloba grows in upland grassland or *Brachystegia* woodland up to about 2 750 m. It may be seen in flower in January in the Kigogo Forest at Mufindi and on the Umalila plateau at Ilembo and in December in the Lupa Forest on the Chunya escarpment.

From Tanzania its distribution extends southwards to South Africa. The Wamalila call this and related species 'insevelakwale'.

BALSAMINACEAE
THE BALSAM FAMILY

Plates 17 and 18

A large family consisting of two genera only one of which, *Impatiens,* occurs in Africa. Herbaceous perennial or less commonly annual plants with succulent brittle stems. Some species have creeping rhizomes whilst a few, such as *I. tinctoria,* arise from a large tuber. Leaves mostly thin-textured, spirally arranged or less frequently opposite, bearing in their axils solitary flowers or fascicles or racemes of flowers which are frequently large and showy and have a spur-like nectary. Fruits quite characteristic, burst explosively when ripe to release the seeds hence their colloquial name of 'touch-me-nots'. They are also collectively referred to as 'balsams' whilst one species *I. walleriana* which is common in eastern Tanzania is the popular pot plant called the 'Busy Lizzie'.

About 70 species of *Impatiens* grow in Tanzania amounting to over half the total number of species found in Africa. Many of the Tanzanian species are very restricted in distribution and some such as *I. flammea* and *I. leedalii* are restricted to single mountain ranges.

Impatiens assurgens Baker [17A]

Several species of balsam or touch-me-not of the genus *Impatiens* are found in the Southern Highlands. These are mostly plants found on the margins or in clearings in the montane forests where they may carpet the ground and produce vivid splashes of colour when in full flower. A few species are however found only in the marshes and grasslands and one of the daintiest of these is *I. assurgens.*

Its simple or few-branched fleshy stems are 10 to 60 cm tall and its opposite leaves are carried towards the apex of each shoot. The leaves may be either glabrous or more or less hairy and are 1.5 to 6.5 cm long. Its flowers are borne either singly or in twos or threes in the upper leaf axils and vary from white through shades of pink to rose-purple with an orange spot at the base of each side-petal. The upper petal is 8 to 12 mm and the side petals 15 to 28 mm long whilst the small purplish or red spur varies from 3 to 10 mm in length.

It is widely distributed in South-central Africa from southern Tanzania and Mozambique across to Angola. In the Southern Highlands it is widespread, for example, on the Mbeya Range and in the marshes around Mbozi. However, we shall always associate it with the swamps of the Hakafilo River near Uwemba in Ubena country. There, large areas of swampy grassland are found on the slopes leading down to the river and, in January and February, *I. assurgens* can be found in full flower growing with *Gerardiina angolensis* and *Impatiens polyantha.*

Plate 17. A, *Impatiens assurgens;* B, *I.austrotanzanica;* C, *I.confusa;* D, *I.cribbii;* E, *I.gomphophylla;* F, *I.leedalii;* G, *I.polyantha;* H, *I.volkensii*

Impatiens austro-tanzanica Grey-Wilson [17B]
I.austrotanzanica, readily distinguished by its dark red cornet-shaped flowers, is a speciality of the mountains south of Mbeya and also the Misuku Hills of neighbouring Malawi.

It is a tall plant up to 1.3 m in height with the leaves arranged in a spiral on the stems. The leaves are slightly hairy and from 3 to 13 cm long. Between one and four flowers, 2.5 to 4 cm long, are produced at each node. These are commonly deep red with a white central column but occasionally pink- or even white-flowered plants can be seen.

I.austrotanzanica grows in shady places and also on the margins of montane forest patches and along rivers and by paths between 1 600 and 2 400 m flowering between October and May. It is common in places in the Poroto Mts, for example, by the village of Isongole on the way to Ngosi Mt, on Mbogo Mt in Umalila and on Rungwe Mt.

In some places in the Poroto Mts forest clearance has brought this species into contact with the grassland species *I.gomphophylla* and *I.rubromaculata* and it then hybridises freely with both species. The hybrids generally are intermediate between the parent species in flower shape and colour.

Impatiens confusa Grey-Wilson [IA & 17C]
A sparingly branched straggling herb up to 1.5 m tall, often rooting at the lower nodes. Three to four glabrous leaves, 13 cm long and 7 cm broad but commonly smaller, are carried at each node on long stalks up to 12.5 cm long. The pink flowers, about 3 cm across, are borne singly at the upper nodes. The bases of the side-petals are darker purplish pink and each of the lower lobes has a yellow spot at its base. The flowers have a filiform spur, 3 to 5.5 cm long, at their base.

I.confusa is more widespread than was at first appreciated but everywhere it is local. The original collections were made on Rungwe Mt, in the Poroto Mts and on Ngosi Mt. It grows in deep shade in the montane forest, in the latter locality in bamboo thickets at between 2 000 and 2 180 m. Varieties of it have recently been found in the Mufindi area at about 1 900 m growing by streams in montane forest and again near Sali, south of Mahenge, where it was found growing in a swamp in montane forest at about 1 400 m.

Impatiens cribbii (Grey-Wilson) Grey-Wilson [17D]
Closely related to *I.gomphophylla* is *I.cribbii,* a balsam restricted to the Kitulo Plateau and Kipengere Mts. It is generally a smaller plant, commonly about 60 to 100 cm tall, but is readily distinguished by its long narrow glabrous leaves, 6 to 13 cm long and 4 to 23 mm broad, and by its more spectacular 2.5 to 3 cm long scarlet-orange flowers with a more inflated spur.

It is common on the Kitulo Farm growing on damp grassy banks and slopes below the farm buildings. In the lower parts of the Plateau it can occasionally be found growing with *I.gomphophylla* but the two do not apparently hybridise.

Impatiens gomphophylla Bak. [17E]

This attractive balsam is probably the most widespread and common species of the Southern Highlands. It often grows in large clumps in grassland. Its stems reach 1.5 m tall and bear finely hairy spirally-arranged leaves from 3 to 9.5 cm long. The flowers borne in bunches of 2 to 6 in the leaf axils resemble those of *I.austrotanzanica* in shape but are distinctly coloured yellow and orange or pale yellow-green and have a more inflated spur. The flowers are between 2 and 2.5 cm long, the constricted part of the spur being 5 to 8 mm long.

I.gomphophylla grows in moist, open or semi-shaded places in grassland or scrub between 1 200 and 2 650 m and flowers almost throughout the year. It can be seen by the roadside on the main road from Iringa to Mbeya at Sao Hill and Makambako. Again it is common on the lower southern slopes of Mbeya Mt and in the Poroto Mts around the village of Isongole and also on the road to the Kitulo Plateau.

On Mbeya Mt it grows with and hybridises with *I.polyantha* whilst in the Poroto Mts it hybridises both with that species and with *I.austrotanzanica.*

I.gomphophylla is widely distributed in south-central Africa in southern Tanzania, Malawi, Zambia and southern Zaire.

Impatiens leedalii Grey-Wilson [17F]

This recently described balsam is a great rarity, having been seen only once on the slopes of Mtorwi Mt on the Kitulo Plateau.

It is a large plant up to 1.2 m tall. The glabrous or slightly hairy spirally arranged leaves are up to 16 cm long and 6.5 cm broad. One to four large flowers borne in the axils of the upper leaves are white or pale pink, the upper petal having a crimson crest and the upper lobe of the side petals having a reddish basal blotch. At the base of each 3.4 to 3.8 cm long lower lobe is a yellow raised boss. The flower which is not unlike that of *I.rosulata* is readily distinguished by its slender 1 to 1.5 cm long spur.

On Mtorwi Mt it grows in montane forest patches at 2 800 to 2 900 m in deep shade and flowers in January and February. It should also be looked for in the forests of the Kipengere and Poroto Mts.

Impatiens polyantha Gilg [17G]

This attractive balsam is a common species of the grasslands, marshes and bushland of our area. It is a plant, 40 to 80 cm tall, often growing in small slumps. The spirally-arranged rarely hairy leaves are 2.5 to 10 cm long and two to five flowers are produced from the leaf-axils. The flowers are white or pink and similar in shape and colouration to those of *I.rosulata* and *I.leedalii* being distinguished by their 4 to 10 mm long spur and 2 to 3 cm long lower petals.

I.polyantha is found only in the Southern Highlands and on the Nyika Plateau in neighbouring Malawi. Although commonly a grassland species it is occasionally found in thickets, on the margins of forest patches and along

PLATE I

streams. A typical locality is on the lower south slopes of Mbeya Mt where it grows with and occasionally hybridises with *I.gomphophylla.* It is also common in the Poroto Mts both on the Tukuyu and on the Kitulo roads. On the latter it occasionally grows with and hybridises with *I.rosulata.*

Impatiens volkensii Warb. [17H]

Rivalling or surpassing *I.tinctoria* in size is the rare *I.volkensii* whose thick succulent stems reach 4 m tall. The large dark green leaves, 11.5 to 28 cm long, are borne in dense spirals on the stems and have hairy veins on the underside. The flowers, relatively inconspicuous for such a large plant, are hidden close to the stem in axillary bunches of 3 to 10. They are white with purple dots towards the base of the side-petals which are 3.4 to 3.9 cm long. The slender basal spur, 8 to 25 mm long, is swollen slightly towards the apex.

It has an unusual distribution in Tanzania for it is known only from Mt Kilimanjaro, where it was first discovered growing at about 2 000 m by streams on the south and west sides, and from Mt Rungwe and the Poroto Mts. We have seen it in flower in February growing in dense stands along the fast-flowing Lufirio River north of Mwakaleli at about 1 800 m and by the Kiwira River at Kanyegere.

Impatiens eryaleia Launert [18A]

This is another balsam of the fairly restricted distribution characteristic of the genus. It is known only from northern Malawi and southwest Tanzania, confined, in fact, to the area around the northern end of Lake Nyasa.

It forms fairly large plants up to 1 m tall, the stems becoming woody and leafless below. The spirally-arranged leaves vary from 4.5 to 13.5 cm long and bear in their axils either single flowers or bunches of a few flowers. The flowers are pale mauve to deep rose in colour with a whitish 3.3 to 4.8 cm long spur. The upper petal varies from 1 to 1.5 cm long while the lower petal may be up to 2.2 cm long.

I.eryaleia grows in moist, usually heavily shaded places in forest and rarely on forest margins from 1400 to 2250 m. We have seen it in flower from January until March in the Bundali Hills above Isoko and also by the Kiwira River.

Impatiens flammea Gilg [18B]

This rare but spectacular balsam has been found in only three localities in the mountains near Tukuyu.

PLATE I

A	B
C	D
	E

A *Impatiens confusa,* Poroto Mts (from Grey-Wilson 1980), p. 85
B *Impatiens rosulata,* Kitulo Plateau (from Grey-Wilson 1980), p. 90
C *Glossostelma spathulatum,* Poroto Mts, p. 103
D *Clematopsis villosa,* near Mbozi, p. 37
E *Swertia welwitschii,* Mbozi, p. 100

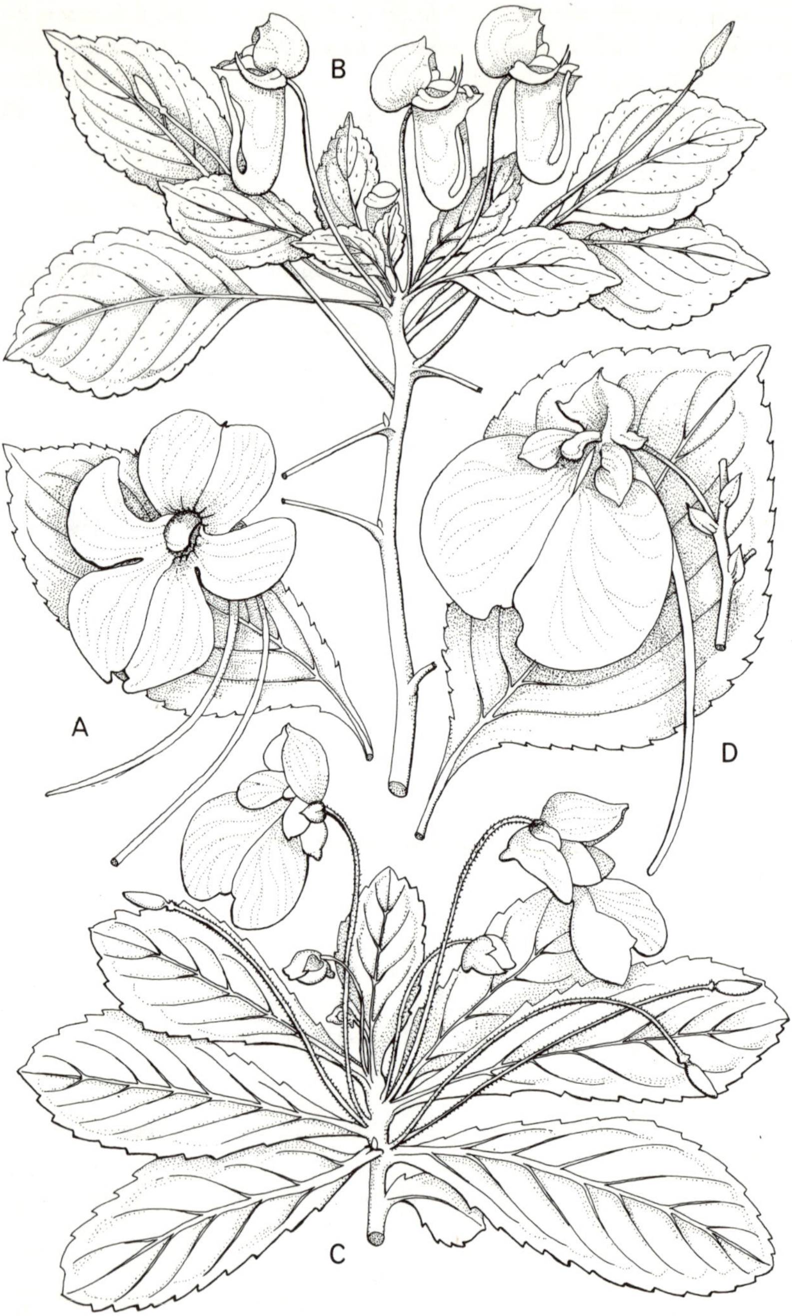

Plate 18. A, *Impatiens eryaleia;* B, *I.flammea;* C, *I.rosulata;* D, *I.tinctoria* subsp. *latifolia*

It is a small plant up to 45 cm tall but is often found growing in dense sheets in the forest. Its spirally-arranged pale green leaves are borne towards the apex of the stem and vary from 2.5 to 10 cm long. They are sparsely hairy on the veins on their lower surface. One to three flowers are borne in each leaf axil on long erect stalks. The flowers themselves point upwards and are flame red or scarlet in colour and about 2.5 cm long.

I.flammea was lost for over half a century after its initial discovery until we recollected it in February 1976 in the Poroto Mts at about 2 200 m altitude. It was growing in bamboo thickets in small colonies on the slopes above the Lufirio River north of Mwakaleli in Unyakyusa. We found it again in the same area near Lake Dhambwe growing with the purple-flowered orchid *Calanthe sylvatica.* The southern end of the area can be reached from Kandete and the northern end, including Lake Dhambwe and Wehe Mt, from the village of Kikondo on the road from Mbeya through Igoma to Kitulo. From Kikondo an easy and beautiful descent through meadows rich in flowers takes one to the Lake. Buffalo-traps, witness to their former presence here, are common everywhere and should be avoided.

Impatiens hydrogetanoides Launert

I.hydrogetanoides is one of the prettiest of the pink-flowered balsams and is also one of the commonest species of *Impatiens* of the higher altitudes in the mountains around Mbeya.

It is an erect herb, often growing in large clumps, with 15 to 70 cm long slender stems and slightly hairy spirally arranged leaves, 2.8 to 13 cm long, clustered towards the stem apex. The flowers, borne in clusters of two or three or singly at the nodes, are attractively coloured rose-pink with a purple mark on each side in the centre. The upper petal is 1 to 1.3 cm long, the large lobe of the lower petal 1.8 to 2.7 cm and the slender spur at the back of the flower 1.8 to 2.5 cm long.

The typical subspecies distinguished by its 7 to 13 cm long leaves and pale pink to lilac flowers is confined to our area and neighbouring parts of northern Malawi and Zambia. It is widespread in the mountains near Mbeya between 2 100 and 2 600 m and can be seen at World's End in the Mbeya Range growing in and on the margins of upland rain forest. It is also common in the Bundali Hills between Ibaba and Ilembo and has been recorded from Imagi Mt just east of Iringa.

Subsp. *kituloensis* Grey-Wilson has a much more restricted distribution, being apparently confined to the Kitulo Plateau and Kipengere Mts between 2 400 and 3 100 m. It is distinguished by its small leaves, 2.8 to 4.5 cm long, and larger deep rose to purple flowers. It is not uncommon growing on the margins of forest patches by the road on the Kitulo Plateau and also on rocky outcrops on Mtorwi and Ishinga Mts.

Impatiens rosulata Grey-Wilson [IB & 18C]
I.rosulata is one of the most prominent plants of the Kitulo Plateau and is also one of the most beautiful. It is commonly a stemless herb with large red-veined dark green leaves, 5 to 18 cm long, in a rosette. The flowers are borne on short stalks in the centre of the rosette standing proud of the leaves and are not unlike those of the much taller *I.polyantha* being attractively coloured white, rose-pink or white and pink. The lateral petals are 2.3 to 2.8 cm long, the large lower lobe having a raised yellow boss towards the base and the smaller upper lobe having a purple patch at the base. The spur at the back of the flower is very short, rarely reaching 2 mm long.

I.rosulata is common all over the Kitulo Plateau growing in open grassy pastures and on banks between 2 100 and 3 100 m. It can be seen in flower from December onwards and the Kinga people call it 'jongojongo'.

Impatiens rubromaculata Warb.
I.rubromaculata is another of the pink or purple species of balsam generally found in the mountain forests. It is a low scrambling to more or less erect perennial herb reaching 1.5 m tall but commonly much shorter. Its spirally arranged leaves are more or less hairy above and below and vary from 1.7 to 11 cm long. The flowers, borne either singly or in pairs in the leaf axils, are pale to deep rose-pink with a dark red blotch on either side in the centre. The upper petal is 0.7 to 1.4 cm long while the lower petals vary from 1.6 to 2.8 cm long.

Four subspecies of *I.rubromaculata* have been recognised and two of these occur in the Southern Highlands. Subsp. *imagiensis* Grey-Wilson known only from Imagi Mt near Iringa has small leaves, 1.7 to 4 cm long, with crenate to dentate margins and lateral sepals less than 5 mm long.

Subsp. *grandiflora* Grey-Wilson has larger leaves, up to 11 cm long, and larger flowers with 6 to 8 mm long lateral sepals. It is known only from the mountains around the top end of Lake Nyasa. It is common in some of the Poroto Mts forests between 1 950 and 2 600 m and can be seen in full flower there in January and February. In places where the forest has been disturbed or cleared it grows and hybridises freely with the dark red-flowered *I.austrotanzanica.*

Of the two other subspecies of *I.rubromaculata,* the typical one is confined to the forests of Mt Kilimanjaro and the other, subsp. *schulziana* (Launert) Grey-Wilson, grows only in the forests of the Nyika Plateau in northern Malawi.

Impatiens tinctoria A.Rich. [18D]
I.tinctoria is one of the largest and most distinctive balsams found in the Southern Highlands, its hollow fleshy stem rising to 2 m tall from a large swollen tuber. The leaves are spirally arranged and vary from 7.5 to 23 cm long. Its 4 to 7.5 cm broad flowers are also quite distinctive and attractive. They are white with red in the centre around the white column. The lower petals are very large,

up to 5.5 cm long and from the back of the flower emerges a long slender spur up to 9.5 cm long (up to 13 cm in some subspp. outside our area) from a conical base.

Five subspecies of this elegant plant are recognised throughout its range from Sudan and Ethiopia southwards to northern Malawi and the Songea District of southern Tanzania. Two of these are recorded from our area. The commonest is subsp. *latifolia* Grey-Wilson, widespread in the mountains around Mbeya and also in neighbouring Malawi growing in damp shady places often in dense forest and by streams from 1 750 to 2 250 m. It has a long flowering period from November until April and can be seen a few kilometres along the Chunya road on the Mbeya Range growing in gulleys in old pine plantations.

Subsp. *songeana* Grey-Wilson has been recorded from southern Tanzania and also from Milo near Njombe. This has smaller flowers and a shorter (5 cm long) thicker spur than subsp. *latifolia.*

The Safwa people call this impressive species 'mbogwi' and the Kinga call it 'mabogoboga'.

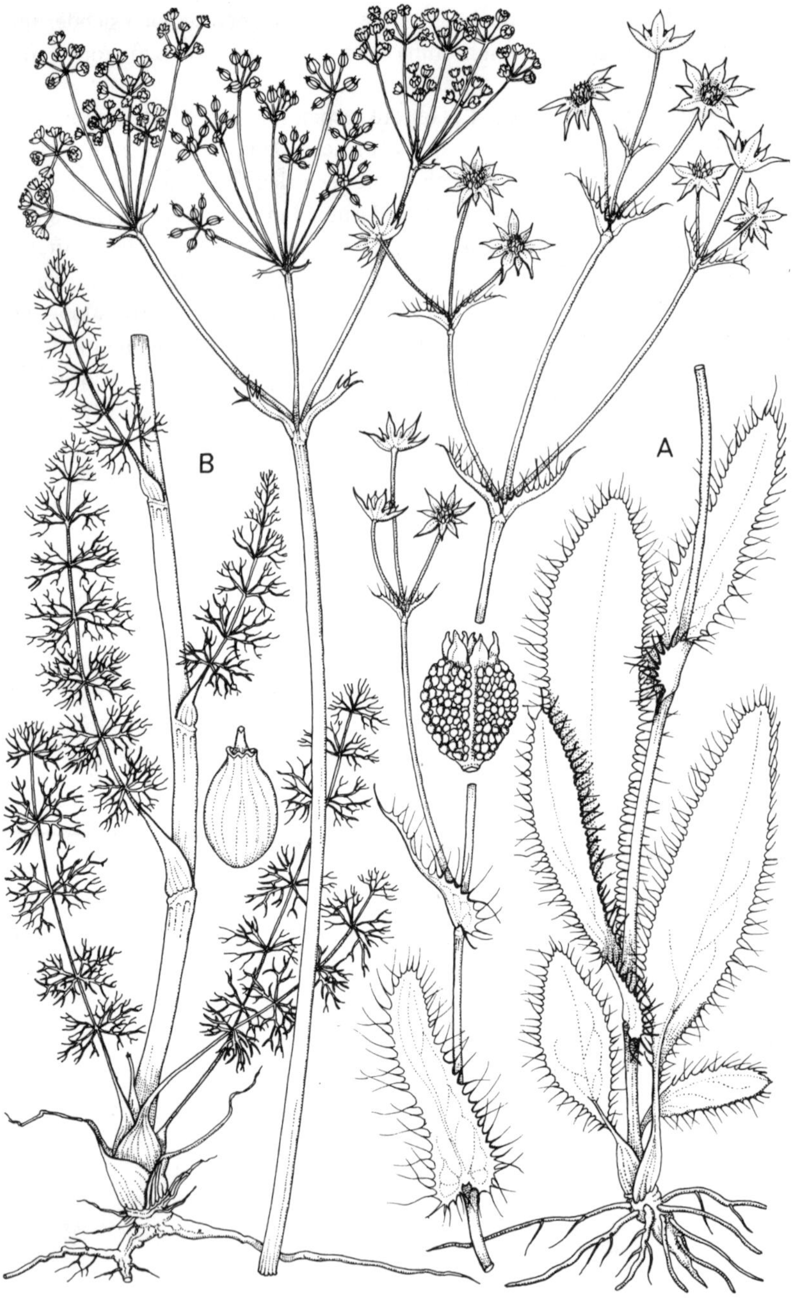

Plate 19. A, *Alepidia swynnertonii;* B, *Frommia ceratophylloides*

UMBELLIFERAE
THE CARROT FAMILY

Plates 19 and 20

Mostly herbs with hollow stems but occasionally shrubby or tree-like (*Steganotaenia* in Africa). Leaves alternate, usually dissected. Inflorescence characteristic, a simple or compound umbel of small usually white or yellow flowers. Umbel characteristic, flat-topped, the individual flower stalks arising from the same point. Flowers bisexual or unisexual, in some umbels the outer flowers sometimes irregular and showier than the inner ones. Each flower consists of a reduced calyx, five petals, five free stamens and an inferior two-locular ovary with two styles. The fruit, a dry schizocarp, is important in the classification of the family.

Several species of the family such as the carrot and the parsnip are important vegetables while others, such as fennel, dill and coriander, are used as herbs for flavouring food.

Alepidia swynnertonii Dumm. [19A]

A relatively slender glabrous herb, 10 to 70 cm tall, with a simple or branched, terete or grooved stem, the leaves variable in size, commonly up to about 10 cm long. Umbels subcapitate, about 1 cm across, with conspicuous pale mauve, greenish cream or white bracts and up to 15 dull brownish flowers. The fruits are verrucose.

This species belongs to a widespread complex and is sometimes included under the aggregate species *A. longifolia* E.Meyer which is widespread in East and South-central Africa southwards to Zimbabwe.

In our area, *A. swynnertonii* is found growing in montane grassland and marshes from 1 700 to about 2 400 m. It is easily seen in flower from January until March by the main road at Iheme south of Iringa at about 1 700 m. Further west, it is common on the Kitulo Plateau and in the Poroto Mts and is also found in Ufipa for example on Nsanga Mt.

Frommia ceratophylloides Wolff [19B]

A herb, up to 1 m tall, growing from a woody rootstock with stout hollow stems 7 to 8 mm in diameter at the purplish base. The more or less glabrous finely divided leaves are mostly basal, up to 10 or 12 cm long, the segments 2 to 5 mm long and 0.5 mm broad. The yellow flowers, about 1.5 mm in diameter, are borne in sparsely branching inflorescences.

A rare plant in Tanzania, apparently confined to Ufipa, where it is found on the Malonje Plateau. It grows there on Nsanga, at about 2 100 m, among large

Plate 20. A, *Heracleum abyssinicum;* B, *Peucedanum muriculatum*

rocks and in rough grassland, flowering in February and March. Elsewhere it is known only from adjacent parts of Zambia and northern Malawi.

Heracleum abyssinicum (Boiss.) Norman [20A]
A perennial or biennial herb up to 1 m tall with solid terete or ridged stems, often purplish at the base, growing from a woody or fleshy rootstock. The simply pinnate leaves are up to 30 cm long and 7 cm broad. The small white flowers are borne in a compound umbel and the fruits are strongly compressed and winged.

It is widespread in eastern Africa from Ethiopia south to Malawi. In our area it can be seen on Mbeya Peak in montane grassland at 2 400 m and also on the Kitulo Plateau from 2 100 m upward. It usually flowers there in January and February.

Peucedanum heracleoides Baker (syn. *P. muriculatum* Welw. ex Hiern) [20B]
A variable herb, 50 to 150 cm tall, with a woody or fleshy taproot, grooved muriculate stems and roughly hairy leaves up to 25 cm long and 18 cm broad. Flowers yellow or purplish-flushed, borne in both terminal and lateral compound umbels up to 10 cm across. Fruits dorso-ventrally compressed, with a papillate-muriculate surface.

Mary Richards collected this species several times in Ufipa, at Malonje Farm growing in rough pasture at 2 250 m, on the east summit of Malonje Mt at 2 550 m and in *Brachystegia* woodland on Kito Mt at 1 800 m. It is also common around Mbeya town at 1 600 m and in Njombe District around Mdapo.

Elsewhere it is widely distributed in south-central Africa from neighbouring Malawi across to Angola.

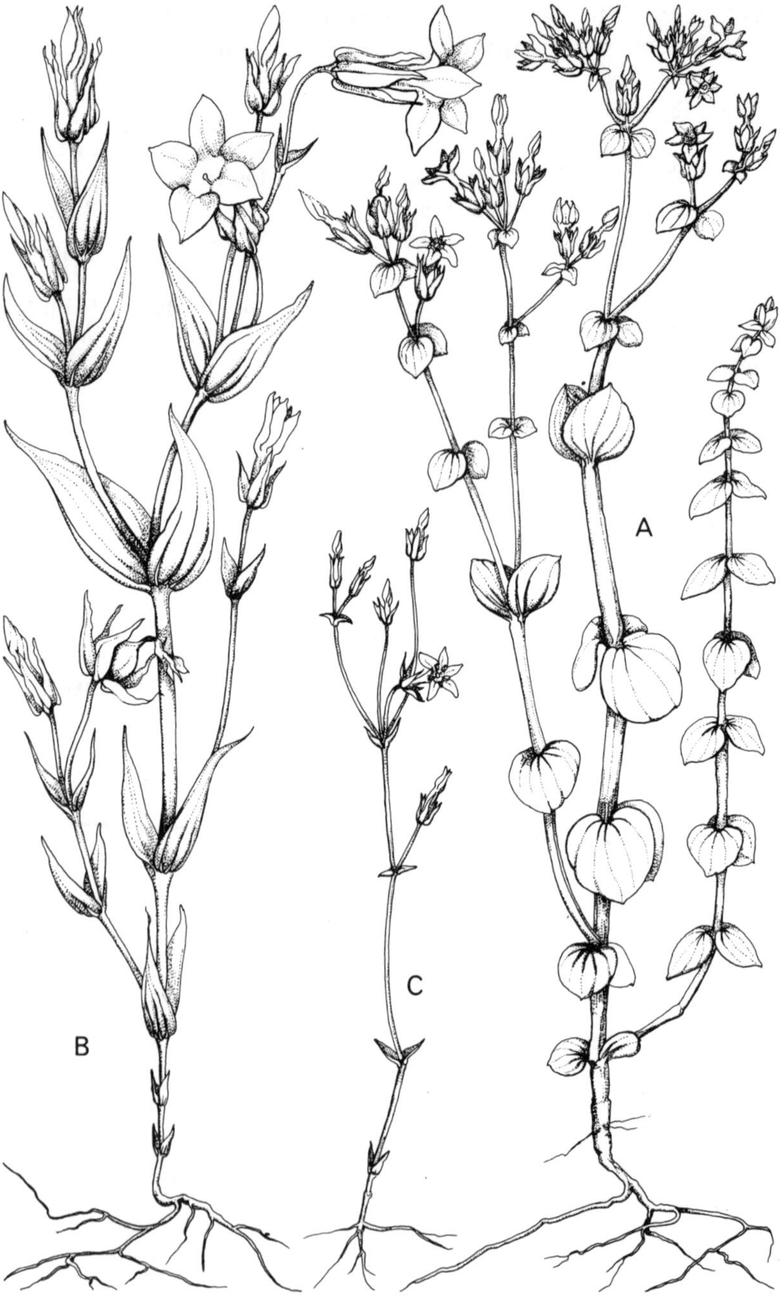

Plate 21. A, *Sebaea brachyphylla;* B, *S.grandis;* C, *S.microphylla*

GENTIANACEAE
THE GENTIAN FAMILY

Plates 21 and 22

Annual or perennial herbs often with a rhizome. Leaves opposite or alternate (in *Swertia*), commonly and characteristically parallel-veined. Flowers bell-shaped, bisexual, with four or five united sepals and a similar number of united petals. Stamens, attached to the corolla tube, equal in number to and alternate with the petals. Ovary superior and fruit commonly a dehiscent capsule.

Many species of the family are attractive plants and those of *Gentiana* are frequently cultivated. The rhizomes of some species (e.g. those of *Swertia* species) contain a bitter principle of medicinal use.

Sebaea brachyphylla Griseb. [21A]

This is a much larger species than *S.microphylla,* up to 45 cm tall with terminal clusters of yellow flowers. Its leaves are ovate, about a centimetre long. It is a mountain flower of wet grassland slopes. A memorable site where we have seen it at its best is on the slopes of the Hakafilo River below Uwemba in Ubena at 2 250 m. Here it grows in association with *Hypericum lalandii, Eriocaulon schimperi, Cyanotis longifolia* and *Otiophora* species. In January this area is alive with flowers and well worth visiting.

Sebaea microphylla (Edgew.) Knobl.

This tiny species flowers late in the wet season in April and May in the mountains of the region. Its slender solitary quadrangular stems, 4 to 15 cm long, grow from a small horizontal white rootstock. It has tiny leaves often only a millimetre long but sometimes up to 6 mm. Its flowers are borne either singly or in a weakly branching inflorescence. Each is erect, yellow and about 7 mm across and typically opens in mid-morning and closes shortly afterwards.

A marvellous area for plants where this tiny species can be seen is on the slopes of the recently active Kiejo volcano south of Mwakaleli. This volcano was last active about 150 years ago when a long tongue of lava flowed down the mountain's side and into the Mbaka valley. Since then plants have slowly recolonised with fig trees, scarlet-flowered *Erythrina* and *Protea welwitschii* being prominent. *S.microphylla,* in association with *Cyanotis barbata, Cynorkis hanningtonii, Streptocarpus* species, *Begonia sutherlandii* and many other plants, can be seen in flower on the rocky volcanic ground in May. The best approach to this area is from the north-west slopes, from which the summit can also easily be reached.

Further east, *S.microphylla* can also be found in the Mufindi area at Lake

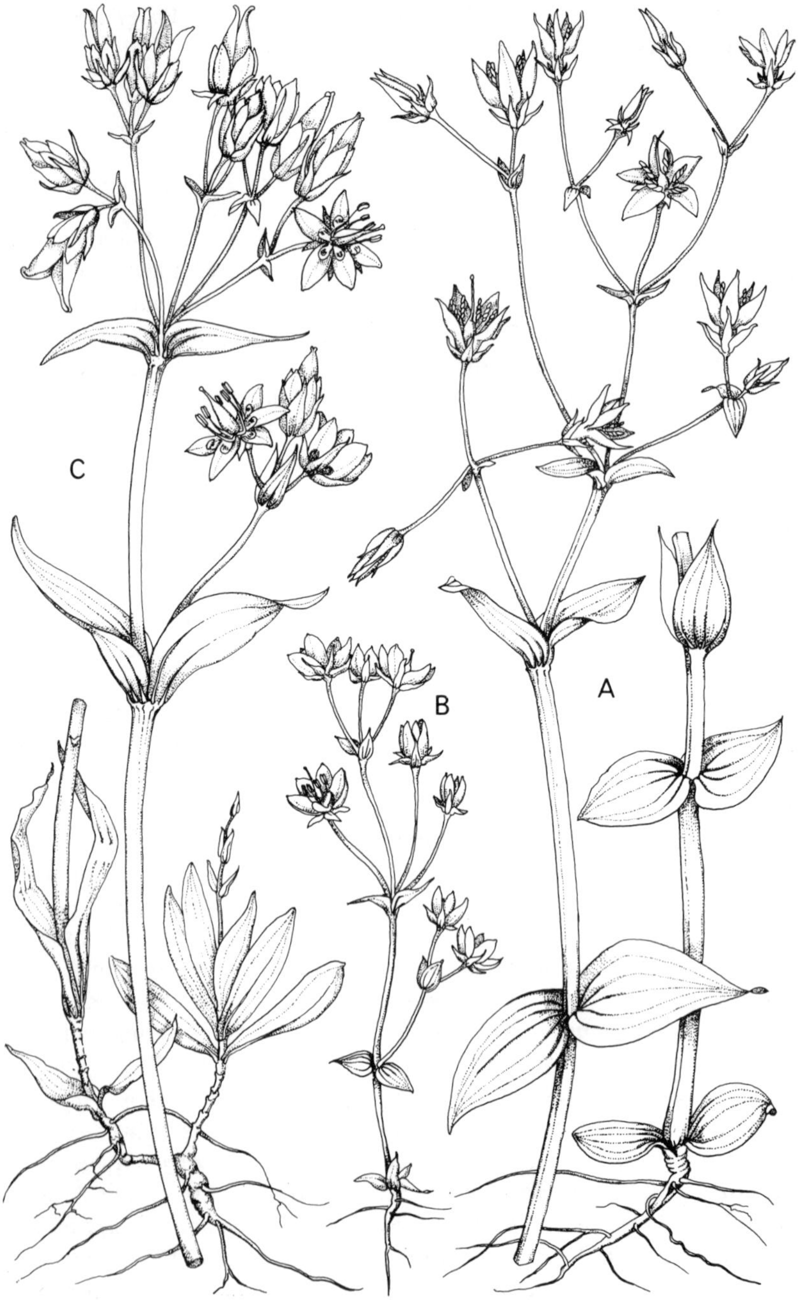

Plate 22. A, *Chironia laxiflora;* B, *Swertia curtoides;* C, *S.kilimandscharica*

Ngwazi where it grows in damp sandy places and in the nearby fields and grassland at 1 900 m.

S.microphylla is at the southern limit of its range in Tanzania but extends to Ethiopia and the Yemen at altitudes up to 2 500 m.

Sebaea grandis (E.Mey.) Steud. [21B]

Several species of *Sebaea,* a genus of yellow- or orange-flowered gentians, grow in the Southern Highlands. Of these, by far the showiest and largest-flowered is *S.grandis.* It is an annual herb, 10 to 35 cm tall, with quadrangular stems arising from a fleshy, creamy white rootstock. Its leaves are up to 3 cm long and 8 mm broad but are rather variable in shape. The pale apricot or rich creamy-yellow flowers, 2 to 2.5 cm long and 2 to 3.5 cm across the lobes, are borne horizontally in branching terminal groups. The calyx-lobes are pale green and up to 1.8 cm long.

S.grandis is widespread in the Southern Highlands and around Sumbawanga. In the eastern part of the Highlands it can be found flowering in March between Matanana and Malangali about 80 km south of Iringa growing in short grassland and in open woodland. It is not uncommon in the Ruaha Park growing in open woodland and also near Lake Kwela in Ufipa at about 1 300 to 1 500 m. Around Mbeya it can be seen in May growing on steep rocky grassy slopes 10 km east of the town.

S.grandis is widespread in Africa from Uganda south to Natal and also in the Cameroun.

Chironia laxiflora Bak. [22A]

This is a flower of May in the Southern Highlands and a very beautiful one too with attractive pink to magenta flowers nearly 3 cm across. The erect or straggling quadrangular stems, branched in their upper part, rise to a metre. Generally the ovate leaves are opposite along the stem, 3 to 5 cm long and 1.5 to 3 cm broad.

It is never a common flower but will be met with quite regularly in grassland or on forest margins at between 1 800 and 2 800 m. On the way to Tukuyu in the Poroto Mts it can be found just past Mporoto at 1 800 m growing in volcanic soil amongst grasses; also on both Mbeya Peak towards the summit, and on Rungwe Mt where it can be seen as late as August on the western slopes growing in secondary bush.

Swertia curtoides Gilg [22B]

Several species of *Swertia,* a genus of gentians with white or lilac flowers, are found in the Southern Highlands. They are pretty plants commonly growing in grasslands and marshes in the mountains. *S.curtoides* is one of the species seen in the montane grasslands from about 2 000 m upwards in our area.

It is an erect branching slender herb, 10 to 40 cm tall, with ovate to lanceolate-ovate leaves up to 4 cm long but commonly much shorter. The stems bear groups of about three white flowers, 1 to 2 cm across, each petal having a green gland at its base.

S.curtoides is confined to the mountains around Mbeya and Njombe. It can be seen in flower from April till August on Rungwe Mt at Ndara at 2 200 m, and also growing in wet valley bottoms in the Kiwira valley at 2 350 m. Again, it grows in the Poroto Mts and on the Kitulo Plateau. The road from Mporoto to the Kitulo Plateau is now relatively good and accessible to most forms of transport. From Igoma, a potato-growing centre, one climbs over volcanic ridges past the village of Kikondo at 2 800 m. Along the roadside superb views over Unyakyusa can be had and in April in the moorlands along the road the white flowers of *S.curtoides* can be seen in their thousands.

Swertia kilimandscharica Engl. [22C]

S.kilimandscharica is one of the loveliest flowers of the Southern Highlands. Its robust erect glabrous stems rise to about a metre tall and are branched near the top. The leaves are 2 to 6.5 cm long and 1 to 3 cm broad. The flowers usually in groups of three at the top are a pale blue or pale mauve and up to 3 cm or so across. Each petal has a gland at the base on its inner side.

We associate it with a superb valley on the east side of Mbeya Peak. If one descends on the path leading back to the road on the north side of the range then turns right down the mountain without ascending the next small peak, a path leads down somewhat indirectly to Lunji Farm. Shortly after leaving the path near the top one drops into a glen which, in April, is full of *S.kilimandscharica.* It can also be seen in flower from August until November on the Kitulo Plateau, at about 2 800 m, growing in moorland that is periodically burnt.

Elsewhere, *S.kilimandscharica* is widespread in the mountains of East Africa from Ethiopia southwards to Tanzania. On Kilimanjaro, where it was originally discovered, it grows in moorland up to about 3 500 m altitude.

Swertia welwitschii Engl. (syn. *S.stellarioides* (Welw.) Fock) [IE]

One of the most attractive and widespread marsh plants of our area is *Swertia welwitschii,* an erect branching annual herb with green or purplish stems 35 to 100 cm tall. Its leaves are borne in opposite pairs, the lower ones reaching 6 cm long and 1.2 cm broad. The inflorescence is commonly well branched and bears many flowers, 8 to 11 mm in diameter. These are white or pale lilac, veined with mauve and with yellow glands at the base of each petal.

It can be seen growing commonly in the wet grassy pans around Mbozi in February and March at 1 500 m and is also to be seen in flower until May in the marshes along the road from Iringa to Mbeya around Mufindi and Sao Hill and also in Ufipa around Kaengesa.

S.welwitschii is a very variable species widely distributed in Africa from Uganda southwards to Natal.

ASCLEPIADACEAE
THE MILKWEED FAMILY

Plates 23 and 24

A large family of herbs, shrubs, climbers or trees often with a milky sap; many exhibiting xerophytic adaptations of stem and leaves. Leaves opposite or whorled, generally simple and entire; minute stipules generally present. Flowers regular and bisexual; the calyx consisting of five partly fused sepals and the corolla of five fused petals. Both or either the corolla and the stamens may bear secondary appendages which form a characteristic corona around the semi-inferior ovary. Pollen generally united into masses rather as in the Orchidaceae. Fruit also characteristic, consisting of a pair of follicles which dehisce to reveal seeds bearing long silky hairs at one end.

Several genera are grown as ornamentals in the tropics notably *Hoya* and *Stephanotis.* Species of the succulent *Stapelia, Huernia, Caralluma* and *Ceropegia* are also grown for their curiosity value.

Ceropegia papillata N.E.Br. [23A]

In a family of unusual flowers none are stranger than the flask-shaped ones of *Ceropegia* species. Several are found in the Southern Highlands and of these *C. papillata* is probably the one most frequently seen. It is a twining plant, 1 to 2.5 m long, climbing up grasses or other herbs in dense bushland or grassland. It grows from a discoid tuber, about 5 cm in diameter, buried about the same depth down in the soil. The slender stems are hairy and purplish with a colourless latex. Its leaves are dull green, tinged with purple, and up to 8 cm long and 4 cm broad. In the axils of the upper leaves clusters of up to 20 flowers are produced in a head 5 to 6.5 cm across. The 1.5 to 2.5 cm long flowers are commonly pale greyish green and either black or darker green towards the apex.

In our area Huber has recognised the variety *cordiloba* distinguished by its cordate leaf bases and a flower with a distended base and lyre-shaped apical lobes. Its fruits are very slender and reach 10 cm long when mature.

We have found *C.papillata* in several places in the Poroto Mts such as at the foot of Ngosi Mt and also on Rungwe Mt. It flowers from January until March at altitudes between 1 600 and 2 250 m. Elsewhere in the Southern Highlands it has been recorded right the way across to Kibengu some 30 km south of Dabaga.

C.papillata is easily confused with the common *C.meyeri-johannis* Engl. a species widespread in East Africa but mostly recorded from lower altitudes. This latter species is most readily distinguished by its larger leaves and tuberous roots.

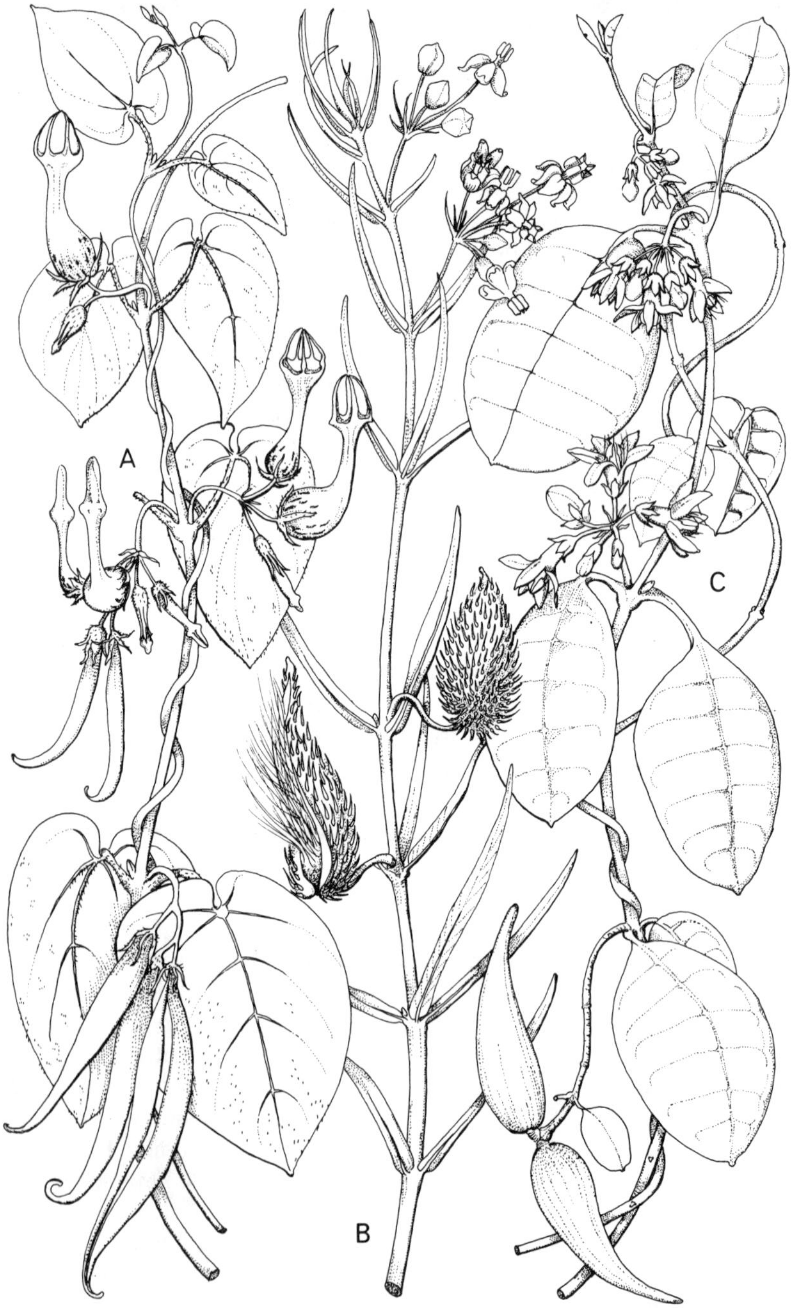

Plate 23. A, *Ceropegia papillata;* B, *Gomphocarpus fruticosus;* C, *Raphionacme welwitschii*

Mary Richards and others have collected several other species from the mountains of our area. In the Uanji-Kipengere ranges she found the lovely *C. chrysochroma* Huber (syn. *Riocreuxia chrysochroma* (Huber) Bullock) a climber which has creamy flowers. *C.schliebenii* Markgraf, with dull brown flowers, is found by the Chimala River at 1 850 m in early January. A distinctive dwarf twining species, generally less than 15 cm tall and bearing only a single but conspicuous flower, grows on the Kitulo Plateau up to about 2 900 m and is perhaps related to *C.adolfii* Schltr. ex Werderm. which can be found in the Kyimbila area.

Glossostelma spathulatum (K.Schum.) Bullock [IC]
A perennial herb, up to 50 cm tall, with leathery leaves and attractive maroon flowers, 4.5 to 6 cm across, with a white central corona, is found between 1 500 and 2 100 m in the *Brachystegia* woodlands of our area. It is also widespread from S.Zaire to Zimbabwe in similar habitats.

Gomphocarpus fruticosus (L.) Aiton f. [23B]
G.fruticosus is a distinctive slender shrub, about 1 to 1.3 m tall, which branches about 60 cm above the ground and has slender leaves to 9 cm long and, at the most, a few mm wide. The flowers are produced in lateral umbels (clusters) about 6 cm across. The individual flowers are greenish cream to yellow with a brown or purplish brown raised central zone called the corona. The corolla is characteristically reflexed to leave the corona clearly exposed as shown in the drawing. The corona is 5-lobed and surrounds a central yellow stigma. The fruit is covered by purple bristles and reaches 5 cm long.

G.fruticosus is uncommon in the Southern Highlands and has been recorded from only two localities, one near Sao Hill growing by the main road at about 1 600 m in a bog in open woodland and flowering in August. We have also seen about eight plants growing in a clump 3 m across in the Umalila hills at 2 200 m also in August. These plants had rather broader leaves than the typical form which is widespread in East Africa.

Raphionacme welwitschii Schlecht. & Rendle (syn. *R.denticulata* N.E.Br.) [23C]
Several species of *Raphioacme* grow in our area but mostly at lower altitudes. However, several can be found in the Mbeya region growing at altitudes up to about 1 600 m.

One of the most interesting of these is *R.welwitschii* a shrubby climber with 60 cm to 1 m tall stems with a milky latex growing from a woody rhizome. The pubescent dark green leaves are large, up to 14 cm long and 8 cm broad. The flowers are produced in 3 cm diameter umbels in the leaf axils along the stem. The individual flowers are relatively small with lobes which are red on the outside and greenish yellow within with a red V-shaped mark at the base of each

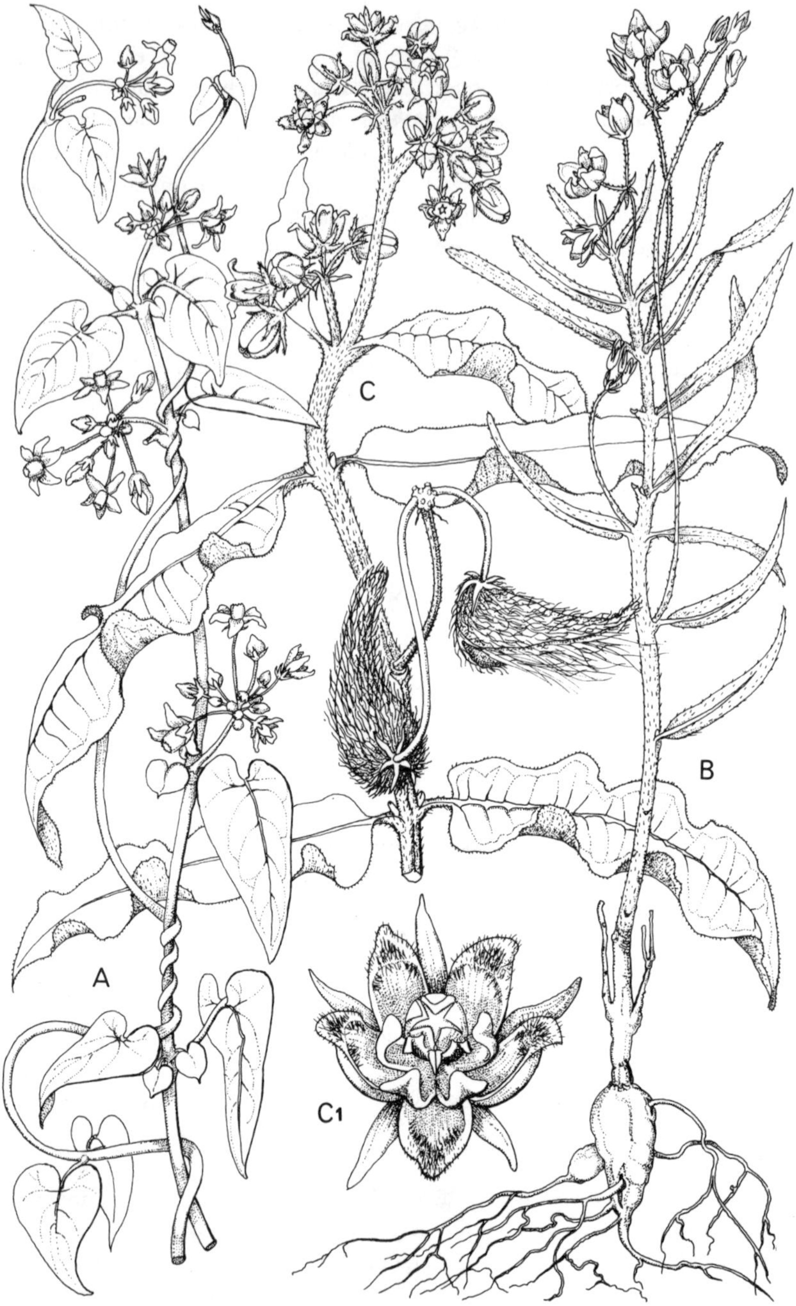

Plate 24. A, *Cynanchum rungwense;* B, *Trachycalymma pulchellum;* C, *Xysmalobium undulatum* (C1, flower)

lobe and a green corona in the centre. The 7 cm long slender fruits are borne in pairs at 180° to each other.

One of the best places to find *R.welwitschii* and allied species such as *R. longituba* E.A.Bruce, *R.madiensis* S.Moore and *R.globosa* K.Schum. is at Songwe just 30 km west of Mbeya. This is an area where a cement factory has recently been set up to utilize the local limestone. This limestone which is, in fact, consolidated tufa has been laid down over a period of thousands of years as the result of hot spring activity which continues to the present day. The calcareous soils at Songwe carry a rich variety of asclepiads which flower in a season which lasts from December until April during which time at least 15 species can be seen. *R.welwitschii* flowers here in December twining up the stalks of grasses in the open bushland near the bat caves which are also a feature of the area.

Cynanchum rungwense Bullock [24A]

This attractive climber is one of about 14 species of *Cynanchum* found in East Africa. It is a slender stemmed plant commonly 2 m or more long with cordate leaves up to 7 cm long and 3.4 cm broad. Its flowers, produced from the leaf axils in umbels about 5 cm across, have greenish or yellowish lobes with a crimson to brownish centre and a yellow-green corona.

It is a common plant in the montane forests of the Southern Highlands. We first saw it, in late November, climbing up a small tree by the roadside on the way from Igoma to Irambo in Usafwa at about 2 250 m. On the opposite side of the valley rises the volcanic cone of Nshovwe densely clothed in forest with giant trees of *Aningeria* and *Entandrophragma* emerging from the canopy.

C.rungwense, as its name suggests, is common on Rungwe Mt and is also widespread in the Poroto Mts, the Mbeya Range and across to Iringa District at altitudes from 2 000 to 2 300 m flowering until March. The Safwa people from around Igoma call it 'ilago'.

Trachycalymma pulchellum (Decne.) Bullock [24B]

The flooded grasslands of the Southern Highlands are wonderful areas for flowers. In East Africa these areas are called 'mbugas' whilst elsewhere in Africa they are referred to as 'dambos' or 'vleis'. They are flat expanses of low-lying land on the gently undulating plateau areas which are seasonally flooded from December until about June. Some of the best are to be seen by any motorist driving from Iringa to Mbeya between Sao Hill and Makambako. Other large groups occur in Mbozi and on the road from Mlowo to Nambinzo. These areas are good hunting grounds for plants including many asclepiads. One of these is the well-named *T.pulchellum* which we have found in fair quantity in the flooded grasslands near Vwawa at 1 550 m in late December.

It is a small plant, 15 to 40 cm tall, its slender pubescent stems arising from a tuberous rhizome. The rather fleshy leaves reach 6 cm long and, in their axils, are borne 4- to 6-flowered inflorescences about 3.5 cm in diameter. The flowers, 1.5 to 2 cm across, are white, flushed with purple or pale pink, with darker veins on their underside and have a green-tipped corona. Its narrow fruits are borne erect and reach 20 cm long.

T.pulchellum is confined in its distribution to south-western Tanzania as far north as Ufipa district where it has been collected at 2 100 m at Chapota growing in woodland amongst rocks.

Xysmalobium undulatum (L.) Aiton f. [24C]

As with most asclepiads *X.undulatum* is never found in great numbers but grows in small scattered clumps and is absent from wide areas of apparently suitable habitat.

It is a stout plant with hairy stems, 1 to 1.3 m long, arising from a large fleshy root. The leaves are large, up to 20 cm long, and are hairy especially on their veins. Both stems and leaves exude a copious milky latex when damaged. The flowers are produced in dense umbels, about 7 cm in diameter, from the leaf axils. The flowers are 1.2 to 1.8 cm across with whitish corolla lobes, conspicuously hairy on their outer surface, and a 5-lobed corona which is dull green, yellow or white and flushed with pink.

It should be looked for in stream valleys in the higher areas from November until April. We first saw it in a small valley near Ilembo at 2 200 m in the Umalila Highlands in January and again at Ndumbi on the Kitulo Plateau at 2 400 m in November. It has also been recorded as far east as Imagi Mt (east of Iringa) growing by a seasonal stream in woodland at 1 800 m.

Elsewhere *X.undulatum* is widespread from Uganda and Kenya south to South Africa.

CONVOLVULACEAE
THE SWEET POTATO FAMILY

Plate 25

Herbs, shrubs, climbers or rarely trees usually with simple alternate entire or variously lobed leaves. Flowers are regular, usually bisexual and sometimes with an involucre of bracts at the base. Parts are in fives with the sepals free and the petals commonly fused into a funnel- or trumpet-shape with the stamens fused to its base. The ovary is superior, commonly of two carpels and the fruit is commonly a dehiscent or indehiscent capsule.

The sweet potato (*Ipomoea batatas*) is the most useful species of the family being widely cultivated throughout the tropics for its large sweet tubers.

Convolvulus farinosus L. [25A]

This is a prostrate or twining herb with slender sparsely pubescent stems. Its cordate or, more commonly, hastate leaves are up to 6.5 cm long and have undulate, serrate or entire margins. The 1- to 6-flowered inflorescences are borne on a peduncle up to 7 cm long. The flowers, commonly white or pinkish white with purple anthers, are about 2 cm in diameter.

It is widespread in East Africa from Uganda and Kenya southwards preferring grassy meadows, damp pastures, old cultivations and rocky places from 1 250 up to 2 800 m. It flowers from late July until December. It is common throughout the Southern Highlands, for example, in the Umalila grasslands at 2 100 m.

C.farinosus is widespread in Africa from the Mediterranean coast to South Africa and is also present in Madagascar and the Mascarene Islands.

A similarly trumpet-flowered bindweed common in our area is *Astripomoea hyoscyamoides* (Vatke) Verdcourt, which prefers the drier areas up to 1 200 m altitude and produces its large white flowers late in the rains or early in the dry season.

Ipomoea alpina Rendle [25B]

This slender herb produces several erect or semi-prostrate stems, up to 25 cm long, from a carrot-shaped tuber and characteristically is covered on its stems and foliage by orange-brown bristle-like hairs. The shortly stalked leaves vary from 2.5 to 7.5 cm long and are up to 6 mm broad. The trumpet-shaped flowers, up to 3 cm in diameter and 1.8 to 3.7 cm long, are borne singly in the upper leaf-axils and are white, pale yellow or orange-yellow with yellowish green venation.

It is widespread in the Highlands from Dabaga in the east to the Mbeya and

Plate 25. A, *Convolvulus farinosus;* B, *Ipomoea alpina;* C, *I. fulvicaulis*

Sumbawanga areas in the west. At Iviji in northern Ulambya it flowers in January on stony ground which appears to be its most suitable habitat. Near Sumbawanga, it can be seen in flower in December growing at the edge of woodland by the Kaka River Falls. Elsewhere in the area, it flowers well into March.

This species is also found in adjacent parts of southern Zaire, Malawi and Zambia.

Numerous other species of *Ipomoea* grow in the Southern Highlands, the best-known of which must certainly be the sweet potato (*I. batatas* (L.) Lam.) originally from South America. Called 'imbatata' around Mbeya, it has large edible tubers. The native species *I. involucrata* Beauv. and *I. wightii* (Wall.) Choisy which are useful for making cord are called 'isintu' by the people around Mbeya.

Ipomoea fulvicaulis (Hochst. ex Choisy) Hallier f. [25C]
This species has very slender twining or prostrate stems, 50 cm or more long, produced from a woody rootstock. The shortly stalked hastate or cordate leaves are up to 5 cm long and covered with golden hairs on both surfaces. The flowers, up to 3 cm long and 2.5 cm in diameter, are attractively coloured pink, rose-lilac or bright mauve.

It is common in the grasslands of the Southern Highlands above about 1 500 m particularly where the grassland has been recently burnt. It can be seen in November by the main Iringa to Mbeya road some 65 km west of Iringa at about 1 900 m growing in grassland and in old cultivations.

I. fulvicaulis is widespread in Africa from Ethiopia south to Zimbabwe and several varieties have been recognised. The plants recorded from our area and from south-central Africa have been assigned to var. *asperifolia* (Hallier f.) Verdc.

Over 30 species of *Ipomoea* have been recorded from the Southern Highlands and it is scarcely surprising that they are a common sight on any roadside.

Plate 26. A, *Solanum indicum;* B, *Cynoglossum coeruleum;* C, *Lithospermum afromontanum*

SOLANACEAE
THE POTATO FAMILY

Plate 26A

Herbs, shrubs or rarely trees with entire or dissected alternate leaves lacking stipules. Flowers bisexual, usually regular, with their parts in fives. Sepals and petals partly fused and commonly the stamens forming a column around the superior ovary and elongate style. Fruit is usually a berry containing many seeds.

This is an economically important family which includes the potato (*Solanum tuberosum*), the tomato (*Lycopersicum lycopersicon*), the egg-plant (*Solanum melongena*), peppers (*Capsicum* species), tobacco (*Nicotiana* species) and many plants with medicinal properties such as deadly nightshade (*Atropa belladonna*) and *Datura* species.

Solanum incanum L. [26A]

This is one of several species of the genus *Solanum,* which includes the potato and eggplant found in the Southern Highlands. All have similar star-shaped flowers with the anthers in a central column around the stigma and berry-like fruits such as are found in the closely related tomato.

S.incanum is a shrub or subshrub to over 1 m tall with softly hairy branching stems. The leaves are variable in size, up to 14 cm long and 8.5 cm broad, but are often smaller and are covered with soft white, brown or purplish hairs. The inflorescence is likewise softly hairy and 2 to 6 cm long with several white, pale blue or pale mauve flowers, about 8 mm across, each with central column of dull yellow anthers in the centre. The pea-sized fruits, at first green, turn orange or red as they mature.

S.incanum is a widespread plant on the margins of mountain forests throughout tropical Africa. It is equally common in the Southern Highlands, at altitudes from 1 800-2 300 m, from Kibengu, south of Dabaga, in the east across to the Mbeya range and the Nsanga Forest in Ufipa in the west. It commonly can be found in flower from February until July in the area.

BORAGINACEAE
THE FORGET-ME-NOT FAMILY

Plates 26B and C

Annual or perennial herbs, shrubs or trees often with roughly hairy stems, leaves and inflorescences. Leaves alternate, simple and usually entire. The inflorescence is quite characteristic consisting of one or more scorpioid or spiral cymes. Flowers regular in most genera, usually bisexual although female flowers are not infrequent. Sepals five, free or connate; corolla five-lobed, flat to bell-shaped, with five stamens inserted on it. Ovary superior, consisting of two fused carpels. The fruit commonly consists of two or four nutlets.

Several genera such as the forget-me-not (*Myosotis*) are cultivated as ornamentals while others such as borage (*Borago officinalis*) are used as culinary herbs.

Cynoglossum coeruleum Hochst. [26B]

A perennial herb, up to 1.2 m tall, covered by bristly white hairs on all its parts. Leaves prominently longitudinally 3-veined, up to 16 cm long but usually much shorter. Flowers tiny, pink, blue or rarely white, about 8 mm in diameter. The seeds are covered with barbed spines.

It is widespread in the marshes and grasslands between Iringa and Mbeya between 1 350 and 2 600 m. At Mufindi, it grows by the shore of Lake Ngwazi flowering in August and can be seen at the same time of the year on Mbeya Mt growing by streams in grassland at 2 600 m.

Other species of *Cynoglossum* also found in our area are *C.lanceolatum*, collected by Mary Richards around the Kiwira fishing camp and at Matamba in Uanji, and *C.geometricum* found on the Mbeya Range.

Lithospermum afromontanum H.Weim. [26C]

A bushy plant, up to 1.2 m tall, with stems rather woody below. Leaves, scabrid, rather glossy, 4 to 6.5 cm long, 0.6 to 1.5 cm broad, are rather pale green on the lower surface. Flowers small, yellow with a darker centre, 8 mm in diameter and fruits shiny, white, about 2 mm in diameter.

It flowers in the wet season from November onwards in the upland grasslands of the Southern Highlands. On the Kitulo Plateau, it grows commonly by rocky outcrops at about 2 400 m. It can also readily be seen in flower on Mbeya Peak in January at about the same altitude. Elsewhere in Tanzania, it is common in the mountains of the north and north-east. On Kilimanjaro, it has been collected at over 3 500 m.

Herbs or shrubs with stems which are characteristically square in cross-section, many of the species having glandular hairs which emit an aromatic fragrance. Leaves simple, opposite and decussate, without stipules. Flowers insect- or bird-pollinated, bilaterally symmetrical and mostly bisexual. Sepals five, fused to form a bell-shaped calyx; petals five, fused; stamens two or four, attached to corolla tube. Ovary superior, the style arising from the base of and between the lobes of the ovary. The fruits consist of four one-seeded achene-like nutlets.

Many species are used as herbs in cooking such as thyme (*Thymus* species) and mint (*Mentha* species). Others such as *Coleus* are widely cultivated for their ornamental value.

Acrocephalus speciosus E.A.Bruce [27A]

This plant, unmistakable because of the dense soft hairs on all its parts, is a small herb with quadrangular stems up to 30 cm tall emerging from a woody rootstock. Its larger leaves, toothed on their margins, reach 10 cm long and 3 cm broad. The flowers, borne at the end of axillary stems a few centimetres long, are congregated in a dense globose head 1 to 2 cm in diameter. Each dull purplish-blue flower is less than 1 cm long with the corolla terminating in four lobes bearing conspicuous tufts of hair. The four exserted white filaments bear black anthers.

It is probably quite widespread in the Southern Highlands having been recorded from Mufindi, Mbeya Mt and also the Umalila region. The figured specimen was collected by Mary Richards at Chapota in Ufipa in March 1957 at 1 600 m altitude. Chapota lies 40 km WSW of Sumbawanga on the edge of the Ipeta swamp and she made extensive collections thereabouts.

A.callianthus is another common *Acrocephalus* species of the mountains which we have seen at Ilembo in May and June and in the Kipengere Mts in January.

Becium cameronii (Bak.) Cribb comb. nov. (Basionym: *Orthosiphon cameronii* Bak. in Kew Bull. 1895:72 (1895). Type: Tanzania, Lake Tanganyika, *Carson* 81 (K, holo.)) [27B]

The mountain commonly called Loleza Peak by the people of Mbeya but more correctly called Kaluwe (c.2 800 m) looms like a giant over the town and many residents and visitors have been tempted to climb its inviting grassy

Plate 27. A, *Acrocephalus speciosus;* B, *Becium cameronii;* C, *Leonotis mollissima*

slopes. The lower slopes are now cultivated and are being washed down into the town but, once above these fields, the climb over the short grass-covered slopes can be savoured. In December, when the rains have begun, squalls of rain come in from the Usangu flats to the east but then a great variety of plants burst into flower. Just above the cultivated area at about 1 900 m grows the smelly *Becium cameronii.* It is a small herb, 25 to 45 cm tall, growing from a stout woody rootstock. Its opposite dark green leaves are up to 4.5 cm long and 1.6 cm broad. At each node there are four white, lilac or pink flowers, each with petals up to 1.8 cm long and exserted styles and stamens 2 to 3 cm long. The calyx is distinctively red, chocolate and green with minute toothing along the margin.

B.cameronii is not uncommon around Mbeya and Sumbawanga and is frequently met with in montane grassland such as that around Chapota. It is called 'umgalijembe' by the Wanyiha people.

Leonotis mollissima Guerke [27C]

The genus *Leonotis* contains some of the showiest and most widespread flowers of the mountains and plains of the area. The majority of them have reddish, orange or yellow flowers and, flowering as they do from June onwards, often give colour to an otherwise dry landscape. The two common species are *L. nepetifolia,* which occurs throughout the lower levels, and *L.mollissima,* illustrated here, from the mountains. The Wakinga people call these plants 'inyunyu', the Wafipa 'nakafundo' and the Wanyiha 'infwonfwo'.

L.mollissima is a large plant, commonly about 2 m tall, growing in large clumps with shortly hairy stems and leaves up to about 10 cm long and densely felted beneath. Dense heads of flowers, 6 to 8 cm across, are produced from the upper nodes. The flowers, exserted 2 to 3 cm from the bracts, are reddish-buff or orange-red and pale cream. The upper lip of the flower is softly hairy but the lower lip is glabrous. Each flower has four stamens, the uppermost longest and arching under the upper lip of the flower on the bristly hairs of which pollen is deposited ready to be brushed on to the sunbirds which frequent these plants. Presumably these are the pollinators of this species.

It commonly grows in bushland or in the open areas in woodland. Plants we have seen near Sheyo in the western Bundali Hills were of tree-size, up to 4 m tall.

The rarer *L.goetzei* which flowers in August is found in the Poroto Mts.

Plate 28. A, *Bartsia abyssinica;* B, *B.nyikensis;* C, *Buchnera crassifolia;* D, *Buchnera pulchra;* E, *B.quadrifaria;* F, *Craterostigma goetzei;* G, *Sopubia ramosa*

SCROPHULARIACEAE
THE FIGWORT FAMILY

Plates 28 and 29

Herbs or rarely shrubs, lianas or trees with alternate or opposite leaves lacking stipules. Inflorescence commonly a raceme or cyme; the flowers irregular and bisexual, often markedly spurred. Calyx five-lobed as is the corolla which is often two-lipped. Stamens four, attached to the petals, two longer than the others. There is a nectar-secreting disc and the ovary is superior. Fruit usually a dehiscent dry capsule.

A family with many ornamental members notably in such genera as *Hebe, Veronica, Verbascum, Antirrhinum, Mimulus* and *Calceolaria.* The Eurasian foxglove, *Digitalis purpurea,* is the source of the important drug digitalin.

Bartsia abyssinica Benth. [28A]

A semiparasitic annual herb with pubescent, erect stems, 19 to 50 cm tall, and pubescent leaves up to about 1.8 cm long and 0.7 cm broad. Flowers pink or lilac, about 1.1 cm long.

Plants in our area are referable to var. *petitiana* (A.Rich.) O.Hedb. It is not infrequent on the Kitulo Plateau where it grows in grassland from about 2 100 to 2 500 m. It is widespread in East Africa south to Malawi and Zambia.

The typical variety is found in Ethiopia and Kenya.

Bartsia nyikensis R.E.Fries [28B]

Closely related to the previous species and considered by some merely a variety, it is a larger scrambling plant with brittle branching stems 60 to 200 cm tall. Leaves generally narrower, up to 2.8 cm long and 2 to 6 mm broad. White or pale pink flowers reach 1.2 cm long and commonly have pinkish anthers.

It can be seen on Mt Rungwe around the lower fishing camp at 1 800 m and is widespread on the Kitulo Plateau up to about 2 800 m. It grows in scrub or on the margins of forest or bamboo clumps and flowers from May through until January. It also grows in neighbouring Malawi and Zambia.

Buchnera crassifolia Engl. [28C]

Usually known in our area under the later synonym *B.rungwensis* Engl. It is a tufted perennial herb up to about 30 cm tall, growing from a woody rootstock and drying black. Leaves sparsely hairy, 2 to 3.5 cm long and 4 to 10 mm broad. Inflorescences about 4 cm in diameter with several lilac, purplish or rich mauve flowers, 1.2 to 2 cm wide.

It grows on the west and north sides of Mt Rungwe on the rocky slopes between 2 300 and 2 700 m amongst *Protea* and *Philippia* scrub. In the Poroto Mts and on the Kitulo Plateau it grows in similar situations from about 1 900 m upwards and can be seen in flower from January until about May.

It is distributed as far south in Africa as Zimbabwe.

Buchnera pulchra Skan & S.Moore [28D]

An attractive perennial herb with dull green, roughly hairy, stiffly erect stems, 40 to 100 cm tall, which branch above. The purple-veined leaves, up to 4.5 cm long and 1 cm broad, are also roughly hairy. Flower-heads compact, terminal, about 3 cm in diameter. Flowers light purple to violet with a white centre, 8 to 11 mm in diameter and having a green calyx tipped with purple.

It is frequent throughout the grasslands of the Southern Highlands. For example, it can be seen on Mbeya Peak at between 2 200 and 2 600 m, near Sao Hill at about 2 000 m and around Mbozi at about 1 400 m. It usually flowers in the area from January until about May.

Elsewhere it is found in Uganda, Zaire, Rwanda, Burundi and south to Zimbabwe.

Buchnera quadrifaria Bak. [28E]

An annual herb, 20 to 45 cm tall, with slender unbranched or weakly branched four-angled stems and leaves up to 5 cm long and 0.5 to 1 mm broad. Flowers lilac to mauve, up to 1.5 cm long, often with white margins.

Around Dabaga and at Holuba in Iringa District, it grows in grassland with scattered trees from 1 800 to 2 100 m. Further west, it can be seen on the Kitulo Plateau, in the Bundali Hills and in Ufipa in similar habitats. It flowers from January until about March.

It can also be found in neighbouring parts of Malawi, Zambia, southern Zaire and Burundi.

Craterostigma goetzei Engl. [28F]

An annual herb with creeping or trailing stems up to about 50 cm long and opposite leaves 1 to 1.9 cm long and 0.8 to 1.4 cm broad. Flowers deep or dark royal blue, marked with white and mauve, reaching 1.8 cm long. Calyx pale green marked with faint blue streaks.

It commonly grows in damp peaty soil by streams and marshes in upland grassland such as at Kilolo, some 40 km south east of Iringa, and around Njombe. It flowers there from about December until February. In southern Zaire, Malawi and Zambia it is not uncommon in similar situations from about 1 250 to 2 350 m altitude.

Sopubia ramosa (Hochst.) Hochst. [28G]

A pretty annual herb with stiff, viscidly hairy, reddish stems up to 1 m tall growing from a woody rootstock and branching near the base. Leaves characteristically drying black, 2.5 to 3.5 cm long, about 2 mm wide. Flowers, pink or rose-purple with a darker centre, 1.5 to 2 cm in diameter.

It is widespread in Africa from Ethiopia and West Africa southwards. We have seen it growing commonly in the grasslands on the Kitulo Plateau around the Kitulo Farm itself. It is also to be seen around Sao Hill and Lake Ngwazi and on Mt Rungwe flowering from about February until May. Although common in the montane grasslands, it can also be found at lower altitudes down to about 1 350 m growing in open woodland in the Ruaha National Park.

Cycnium adonense E.Mey. ex Benth. [29A]

One of the loveliest flowers to be seen in the grasslands of the Mbeya Range and elsewhere in the Southern Highlands is the large white-flowered *Cycnium adonense.* It is a creeping or low, compact herb with hairy green stems, 30 to 100 cm long, tinged with purple. The coarsely hairy leaves, up to 9 cm long and 4 cm broad, similarly are usually dark green strongly tinged with purple. The flowers although large, 5 to 7 cm across with the corolla tube 5 to 6.5 cm long, have delicate white petals tinged with pink and sunken green veins towards the centre. When bruised they turn inky blue, hence the common name of the 'ink flower'. The same colour change is also seen in the flowers of the orchids *Calanthe sylvatica* and *Phaius occidentalis.*

Apart from the Mbeya Range, *C.adonense* can be seen in the grasslands around Sumbawanga at 2 100 m and around Njombe at 1 860 m and is also common in the grasslands of neighbouring Malawi. It mostly flowers from November through until February and again at the end of the rains.

Gerardiina angolensis Engl. [29B]

This, the only species in its genus, is a herb up to 65 cm tall which turns black on drying. Leaves numerous, opposite, the lower 8 to 11 cm long and 1.5 to 1.8 cm broad but the upper smaller. Eight to 14 very pretty white to pinkish mauve flowers are arranged oppositely on short stalks at the top of the stem. Corolla five-lobed, about 2 cm long and 3 cm in diameter, subtended by a cup-like calyx about 6 mm long.

It grows in swampy ground at about 2 200 m in the Southern Highlands. In the Kipengere Mts and by the Hakafilo River at Uwemba, Ubena it can be seen in flower in January.

Veronica abyssinica Fresen. [29C]

A creeping or scandent herb with slender branching pubescent stems, up to 90 cm long, which root at the nodes. The sparsely to quite densely pubescent leaves, 2.5 to 4 cm long, are borne on a 1 to 2 cm long stalk. The flowers, varying from 6 to 12 mm in diameter, are bright blue with a central ring of purple around the white eye.

It is a common plant of the wet grasslands in the Southern Highlands above

Plate 29. A, *Cycnium adonense;* B, *Gerardiina angolensis;* C, *Veronica abyssinica*

about 1 900 m. It is particularly frequent in the Poroto Mts where it can be seen in flower from October until February or March, growing by the main road to Tukuyu at between 2 300 and 2 600 m. It is also frequently seen growing in grassland on the track to Ngosi Mt, on Mt Rungwe and on the Kitulo Plateau, often by streams. We have also found it at Mufindi on the forest margins below the Idege headquarters of the Livalonga Tea Estate.

This species is widespread in Africa from Ethiopia south to Zimbabwe.

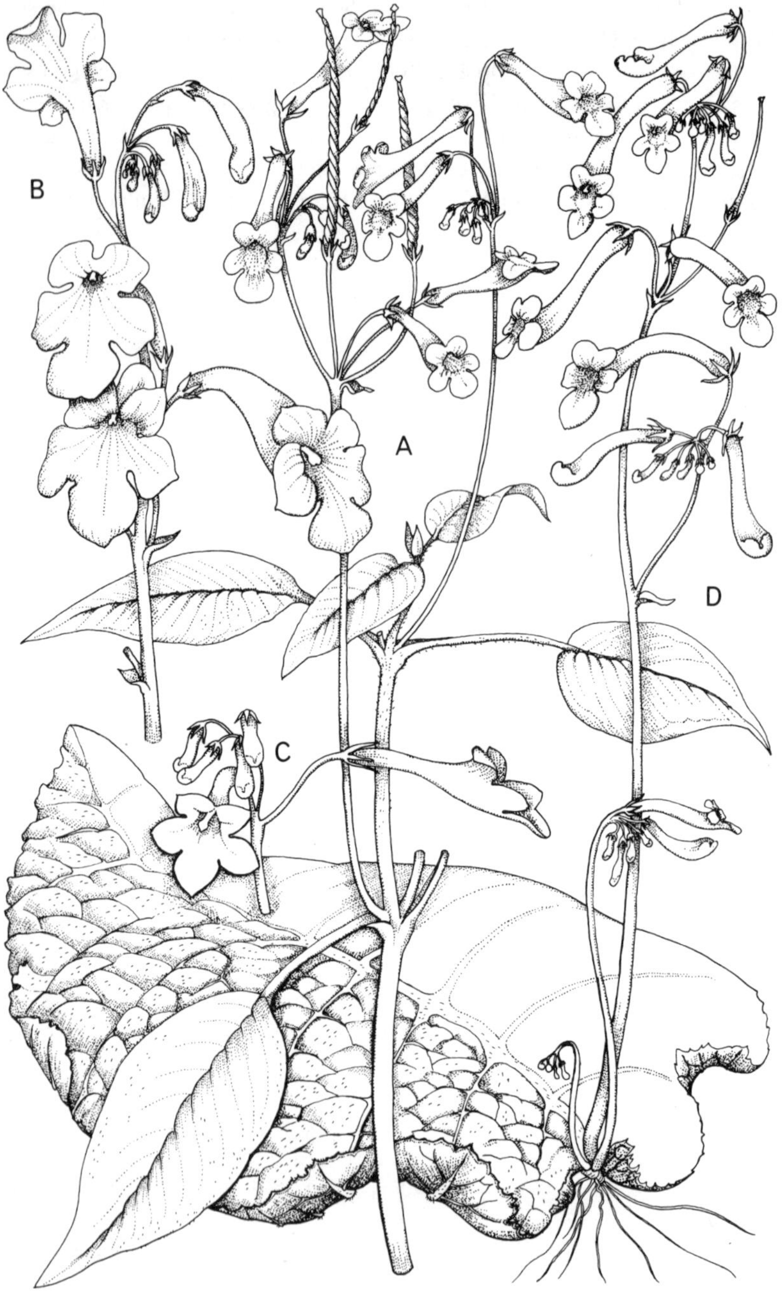

Plate 30. A, *Streptocarpus buchananii;* B, *S.eylesii* subsp. *brevistylis;* C, *S.goetzei;* D, *S. solenanthus*

GESNERIACEAE
THE AFRICAN VIOLET FAMILY

Plate 30

Herbs or shrubs often with rather succulent stems and foliage, the leaves opposite or alternate (or occasionally solitary and basal as in *Steptocarpus*) lacking stipules. Flowers bisexual, irregular, with five sepals united at the base; petals united, two-lipped. The two or four stamens often cohere in pairs. Ovary superior or inferior, developing into rounded or elongate capsules containing many small seeds.

Many species are ornamental and *Saintpaulia, Gloxinia* and *Streptocarpus* are all popular in cultivation. Plants can often be grown from leaf-cuttings. Although *Saintpaulia* has not been found in the Southern Highlands, it could well be searched for on the eastern flanks facing the Uluguru Mts where it does occur.

Streptocarpus buchananii C.B.Clarke [30A]

A straggling plant with a long succulent stem bearing many pairs of opposite, petiolate, roughly hairy leaves, 6 to 11 cm long. It is one of the smallest flowered species found in the Southern Highlands but its flowers, about 2 cm long, are strikingly coloured, deep blue or violet, and many are borne in axillary inflorescences, 25 to 30 cm long.

In our area, it flowers from January to March and can be seen in the forests of the Poroto Mts, Bundali Hills and on Rungwe and Ngosi Mts between about 2 100-2 500 m growing at the base of trees or epiphytically on their lower branches. *S.buchananii* is also found in neighbouring Malawi and Mozambique.

Streptocarpus eylesii S.Moore [30B]

Probably the most spectacular of the Southern Highland species and, unlike the others, often found outside the montane forests for it prefers wet rock faces and stream banks. It has a large densely white-pubescent basal leaf, up to 45 cm long, flat on the surface. One to four erect inflorescences, up to 30 cm long, bearing several showy pale to deep blue-violet or rarely white tubular flowers, up to 4.6 cm long, are produced at the base of the leaf.

Plants in our area and in neighbouring northern Malawi and Zambia are referable to subsp. *brevistylis* Hilliard & B.L.Burtt. It is widespread in the Southern Highlands from Imagi Mt, east of Iringa, across to Sumbawanga and Mbeya at altitudes above 1 800 m. Above Mbeya it can be found growing on the rocky outcrops of Loleza Mt at about 2 000 m right up to the summit of Mbeya Mt at over 2 900 m.

The typical subspecies, differing in its longer style and smaller stigma, is found further south in Africa in Zimbabwe and Malawi.

Streptocarpus goetzei Engl. [30C]

A most attractive species allied to *S.eylesii* and to *S.solenanthus.* Similar to these it has a single, prostrate, relatively narrow and only sparsely pubescent leaf with a crenate margin, up to 28 cm long but commonly rather shorter. Inflorescences arise from the base of the leaf and reach 30 cm long, the tubular flowers being borne at the apex of a pubescent peduncle. The flowers, intermediate in size between *S.eylesii* and *S.solenanthus,* are about 3.6 cm long and a rich blue-violet in colour.

S.goetzei is another species which may be found growing epiphytically on the trunks of forest trees in the area at altitudes between 1 900 and 2 500 m but it is also found on damp banks and by streams. In places, it can frequently be found in colonies flowering between January and March, for example, on Mt Rungwe near the Kiwira Fishing Camp and in the Livingstone Mts Forest Reserve where it is not infrequent in the bamboo zone.

Streptocarpus solenanthus Mansf. [30D]

Widespread in southern Tanzania from the Uluguru Mts across to Mbeya and southwards to the Mozambique border, it is a most distinctive species with a solitary, prostrate, finely pubescent leaf, up to 35 cm or more long, at the base of the plant. One to four erect inflorescences up to 30 cm tall, each bearing several trumpet-shaped flowers at the apex, are produced from the leaf base. The flowers usually white or less frequently tinged with pale violet are about 2.8 cm long.

S.solenanthus commonly grows on damp banks, epiphytically on the lower branches of forest trees or by streams in montane forest. It was first collected by H.J.Schlieben in the forests above Mahenge where it is the only species of the genus to be found. It may be seen also at Mufindi growing on rocks and trees overhanging the streams of the upper escarpment forest.

ACANTHACEAE
THE BLACK-EYED SUSAN FAMILY

Plates 31 and 32

Herbs, shrubs or rarely trees with opposite decussate leaves, lacking stipules. Flowers bisexual, bilaterally symmetrical, solitary, in cymes or in racemes, often enclosed in large petaloid bracts. Sepals and petals four or five, the latter fused usually into a two-lipped corolla; stamens two or four, attached to the petals, sometimes with one or two staminodes. Ovary superior, of two fused carpels. Fruit a bilocular capsule.

A number of species are cultivated as ornamentals including species of *Thunbergia, Crossandra* and *Justicia.* Others are used medicinally.

Dyschoriste trichocalyx Lindau [31A]

A herb, 30 to 100 cm tall, with branching hispid stems growing from a woody rootstock. The leaves are also hairy, up to 5.5 cm long and 2.5 cm broad. Dense masses of sessile whitish to pale mauve flowers, each with four stamens and a hairy calyx, are borne in the upper leaf-axils.

It can be seen in flower in mid-May growing in grassland between Santilya and Isuto in the Umalila Highlands and also north of Majengo near Mbeya. It is widespread in East Africa in similar habitats.

Most of the other species of the genus found in the Southern Highlands are shrubs. One of these, *D.verticillaris,* found in the Mbisi forest near Sumbawanga and elsewhere, has bright mauve flowers with darker markings. Also in the Mbisi forest is the white-flowered *D.tanganyikensis* and both can be seen in flower in August.

Hypoestes aristata (Vahl) Roem. & Schultes [31B]

Commonly a straggling plant, 0.5 to 2 m tall, with pubescent branching stems and elliptic leaves, 4 to 10 cm long and 2 to 5 cm broad. The flowers, borne in whorls in the leaf axils, are white, pink or pale mauve and 2.5 to 3.5 cm long. The lower lip of the flower is entire but the upper is three-lobed and red-spotted on the central lobe.

H.aristata is widespread between 1 700 and 2 200 m in the Highlands and is often found growing on the margins of, or in, forest. It flowers from July until October and can be seen from Imagi Mt in the east across to the Porotos and Mbeya Peak in the west. At Mahenge, it grows on the tops of large calcite outcrops, at about 1 000 m, along with *Sansevieria* and succulent *Euphorbia* species.

It is widespread in the mountains of Uganda and Kenya and throughout South-central Africa.

Plate 31. A, *Dischoriste trichocalyx;* B, *Hypoestes aristata;* C, *Isoglossa lactea;* D, *Justicia nuttii*

Isoglossa lactea Lindau [31C]
Several species of the genus *Isoglossa* are found in the Southern Highlands mostly as undergrowth herbs in the montane forests. One of the commonest is *I.lactea,* figured here, a glabrous herb with erect or ascending stems up to 1.2 m tall. Its lower leaves reach 14 cm long and 6 cm broad but the upper ones are much smaller. White flowers, spotted with red or mauve and 1.5 to 2 cm long, are borne in terminal panicles. The corolla is two-lipped, the upper one being bilobed and the lower one trilobed.

Mary Richards collected the plant illustrated here in May on the Kitulo Plateau. We have also seen it in flower in July on the margins of forest on the northern slopes of Rungwe Mt. In the Poroto Mts, it is found as the undergrowth in bamboo forests at about 2 350 m.

It is widespread elsewhere in East Africa including Kenya and on most of the Tanzanian mountains including Kilimanjaro where it grows up to 2 800 m.

Justicia nuttii C.B.Clarke (syn. *J.goetzii* Lindau) [31D]
Many species of *Justicia* are found in the Southern Highlands and *J.nuttii* is probably the commonest in the higher grassland areas. It is a small pubescent herb with several clustered stems up to 26 cm tall and leaves 2 to 3.7 cm long and about 1.5 cm broad. The flowers which are pure white with a few red markings on the mid-lobe of the lower lip are about 1.2 cm long with a two-lipped corolla, the upper lip covering the stamens being slightly two-lobed at the top while the lower lip is more deeply three-lobed. Two brown stamens, each two-celled, lie on either side of a reddish style.

On the Kitulo Plateau it can be seen in flower from December until March and Mary Richards collected it there at 2 800 m at Kikondo in January 1961. Unfortunately, potato farming has greatly changed the landscape there in recent years. However, *J.nuttii* can still be commonly seen on the Plateau and elsewhere in the Southern Highlands from about 1 500 m upwards in the grasslands around Dabaga and Mufindi across to Ufipa and Mbozi. It is common also in neighbouring parts of Malawi and Zambia.

Other species recorded from our area are *J.calophanoides* collected by Mary Richards in the Kipengere Mts, *J.pinguior* which is widespread throughout and *J.linearispica* collected again by Mary Richards in the Mbisi Forest, Ufipa in August.

Thunbergia gregorii S.Moore [32A]
This, one of the more startling flowers of our area readily recognised by its bright orange flowers and orange pubescence of all its parts, is a perennial trailing herb with hastate leaves, up to 8 cm long and 6 cm broad, borne on a winged stalk about 4 cm long. The trumpet-shaped flowers, 4.5 to 5 cm in diameter, are enclosed by a 3 to 4 cm long calyx at the base.

It is widespread in the Southern Highlands from the Iringa area across to

Plate 32. A, *Thunbergia gregorii;* B, *T.lancifolia;* C, *T.natalensis*

the mountains around Njombe and Mbeya. On a memorable day in November we collected it in the *Brachystegia* woodland on the way up to Masusa Mt from the village of Utilile in Upangwa country. (The views from the top of this mountain over Lake Nyasa spreading to the west and the Njombe Highlands away to the east are panoramic.) It is also common on the Mbeya Range and can be seen in January and February in the grasslands by the roadside on the way from Mbeya to Chunya at about 2 000 m. Elsewhere, it is widespread in the mountains of Tanzania, Kenya and south-central Africa from 1 500 m to about 2 350 m.

The genus *Thunbergia* provides some of the most attractive flowers of the Southern Highlands, the best-known being the ubiquitous *T.alata* popularly called 'Black-eyed Susan' because its yellow flowers have a black throat in their centre. It is called 'kijago-gura' in Swahili and 'ngelengele' in Nyakyusa. *T.kirkiana,* with cream flowers, has been collected by Mary Richards in the Tunduma and Mbozi areas. She also collected the blue-flowered *T.lathyroides* in the latter area whilst from southern Njombe district she recorded *T.cordata* whose white flowers have a yellow throat.

Thunbergia lancifolia T.Anders [32B]

This rather variable species, often found growing in clumps 45 to 100 cm tall, has stems which branch half way up produced from a woody rootstock. The leaves are particularly variable in shape from almost elliptic to linear-lanceolate and from 2.5 to 11 cm long and 0.8 to 4 cm broad. Its flowers, commonly dark purple or royal blue with a yellow throat and about 5.5 cm long and 4.5 to 5.5 cm in diameter, have a glandular hairy calyx about 3 cm long.

We have collected it several times in the Southern Highlands. In the Umalila Highlands it flowers from July until October whilst in the woodlands of the Chimala escarpment it can be seen in flower in January and in the Mbisi Forest in Ufipa in November. It is also found growing in montane grassland at Mufindi by Lake Ngwazi and in similar habitats around Mbozi. It is called 'manunywi' by the Wahehe, 'mwimba' and 'fyofyonta' by the Wafipa and 'sonya' by the Wamalila.

T.lancifolia is widespread in the southern half of Africa, south to South Africa.

Thunbergia natalensis Hook. [IIB & 32C]

Similar to *T.lancifolia* this is an extremely variable species particularly in its habit, leaf shape and flower colour. In our area, it is an erect branching herb to about 100 cm tall, often found growing in small clumps, and with its reddish stems bent at the nodes. Its leaves, hairy below particularly on the veins, are usually 5 to 9 cm long and 4 to 5 cm broad. The flowers, 3.5 to 4.5 cm long and in diameter, are pale blue, lilac-blue or rarely white with an orange or yellow throat.

It is widespread from about 1 800 to 2 400 m altitude in the Southern Highlands growing in grassland which is seasonally burnt. A typical locality is Isela near Ilembo in the Umalila Highlands where it is to be found in full flower in October just before the onset of the rains. Elsewhere, it can be found right through until February or even March.

T.natalensis is, as its name suggests, found as far south as South Africa and its range extends northwards to the Kenya Highlands.

PEDALIACEAE
THE SESAME FAMILY

Plate IIC

Annual or perennial herbs or rarely shrubs with opposite, entire or lobed leaves often possessing glandular hairs but lacking stipules. Flowers solitary or borne in a few-flowered axillary cymes with glands at the base of the flower stalks, bisexual and irregular with a calyx of five fused sepals and a tubular corolla of five fused petals. Stamens mostly four, borne on the petals, two being longer than the others. Ovary superior and developing into a capsule or hooked nut.

This is a small family of which *Sesamum* with about 30 species is the largest genus.

Sesamum angolense Thonn. [IIC]

One of the commonest roadside herbs by the main road from Iringa to Mbeya particularly in the highlands around Sao Hill is the attractive *S.angolense.* It is a large plant, often up to 3 m tall, with simple or branching sparsely hairy stems.

Its leaves are entire and shortly stalked, 2 to 11 cm long and 0.4 to 4 cm broad, the lower surface being white-felted especially on the younger leaves. The showy flowers vary from pink or purple to pale mauve with deeper markings. The pubescent corolla tube is large, from 3.5 to 7 cm long and 2 to 3 cm across. The obscurely 4-angled capsule is rather densely hairy with a somewhat flattened broad apical beak.

S.angolense is widespread in Africa from Uganda and Zaire south to Zimbabwe and Angola, growing in wooded and upland grassland and also as a weed of cultivation and in waste places from 350 up to 2 300 m altitude. The Wahehe call it 'litende', the Wanyiha 'usambwe' and the Wafipa 'isonso'.

Of the four other species of *Sesamum* recorded from East Africa, two also grow in the Southern Highlands. One of these, *S.angustifolium* (Oliv.) Engl., is native and common in waste places and in grassland. It differs from *S.angolense* in having leaves which are glandular below and smaller flowers less than 3.5 cm long. The other, *S.indicum* L., is an introduced species native elsewhere in Africa and also in Asia. It is distinguished by its glabrous stems and long-stalked and divided lower leaves and its seeds are utilised for oil.

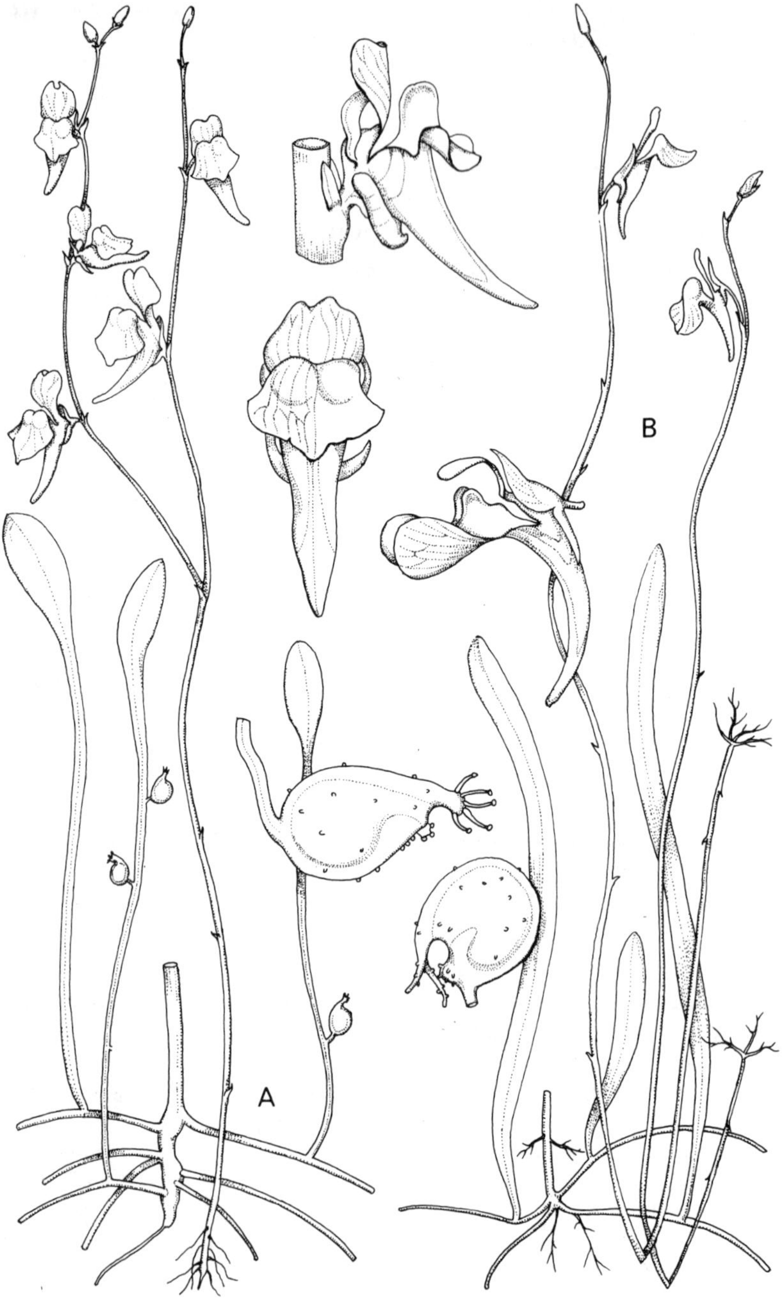

Plate 33. A, *Utricularia firmula;* B, *I.prehensilis*

LENTIBULARIACEAE
THE BLADDERWORT FAMILY

Plate 33

Carnivorous herbs often lacking roots and frequently found in aquatic or moist habitats. Leaves covered with glandular hairs (in *Pinguicula*) or bearing small bladders or traps (*Utricularia*) which secrete digestive enzymes. Small animals are trapped and digested by these providing the plants with nitrogenous compounds.

Flowers irregular, bisexual, with a two- to five-lobed calyx and a five-lobed, two-lipped, often attractively coloured corolla; spurred at the base, bearing two stamens. Ovary superior, producing an indehiscent capsule which opens by two or four valves.

Utricularia firmula Oliv. [33A]

Some 17 species of bladderwort occur in one part or another of the Southern Highlands. These dainty carnivorous plants favour the marshes and wet grasslands which are a feature of mountains here. Many have their feet in water for most of their life and their small bladders catch microscopic animals which are digested to provide nutrients for the plant's growth.

U.firmula is widespread in Tanzania below about 1 800 m altitude. It is a small terrestrial herb up to 36 cm tall but often smaller. Its leaves, from 2 to 20 mm long, are scattered on the stolons. Ten to 20 or so strongly zygomorphic flowers are borne on the scape. The 3.5 to 6 mm long corolla is pale yellow and has a subulate spur at the base of the lower lip.

It may be found growing in the Dabaga Highlands south-east of Iringa where it flowers in February. It is widespread in tropical and southern Africa and is also found in Madagascar.

Utricularia prehensilis E.Mey. [33B]

This is yet another widespread species of bladderwort widely distributed throughout tropical and southern Africa and also in Madagascar. It is a terrestrial species, 10 to 35 cm tall, with numerous leaves up to 10 cm long and 3 mm broad. The 1 to 8 delicate flowers, an attractive yellow in colour, are borne on an erect or more typically twining slender scape.

A typical habitat for *U.prehensilis* is in the boggy ground by the small tributary streams that flow into the Hakafilo River near Uwemba in the Ubena Highlands at 2 000 m where it can be found in flower in early January. Nearby and almost hidden because of its tiny size is the violet and yellowish-flowered *U.welwitschii* Oliv., a species at the northern limit of its distribution in Tanzania. Other plants commonly found growing in association with bladderworts are the sundew, *Drosera madagascariensis,* the fern ally (clubmoss) *Lycopodium affine,* the lovely red swamp orchid *Disa welwitschii* and the daisy *Felicia boehmii.*

U.prehensilis is widespread in boggy places in the area up to 2 100 m altitude.

PLATE II

A	B
C	D

A *Canarina eminii,* Bundali Mts, p. 191
B *Thunbergia natalensis,* World's End, Mbeya Range, p. 129
C *Sesamum angolense,* near Mafinga, Mufindi, p. 131
D *Scadoxux multiflorus,* above Chensema, Uluguru Mts, p. 193

CAMPANULACEAE
THE BELLFLOWER FAMILY

Plates 34 and 35

Mostly herbs but occasionally shrubs or climbers (e.g. *Canarina*). Leaves alternate or sometimes opposite and are always simple. Flowers regular, bisexual, with their parts in fives. Sepals are fused into a tube which is also fused to the semi-inferior ovary. Petals more or less fused commonly into a bell-like corolla. Fruit a capsule.

A family of mostly attractively flowered species notably, in our area, the climbing *Canarina* (see Plate 55B).

Wahlenbergia capillacea (L.f.) A.DC. [34A]

The tropical African specimens of this widespread species are referable to subsp. *tenuior* (Engl.) Thulin, the typical subspecies being confined to South Africa, Lesotho and Swaziland.

The glabrous or ± hairy stems reach 50 cm long and the leaves vary from 5 to 30 mm long. The small blue bell-flowers are 5 to 13 mm long and are lobed for one-third to one-quarter of their length. Subsp. *tenuior* differs from subsp. *capillacea* in the absence of long hairs in the flower tube and by its leaves which are never borne in fascicles.

Mawemapanza, a mountain by the Igoma to Kitulo road in the Poroto Mts, is a marvellous area for flowers. The road passes just below the summit at about 2 850 m and, if the steep bank by the road is climbed, a short walk across grassland leads to the top where a superb view can be obtained providing the mists have lifted. The name of the mountain comes from the Safwa word for white rocks referring to the cliffs of whitish lava on the west side. Here, from April until July, *W.capillacea* will be found hanging on the steep rock faces. Elsewhere in the Southern Highlands it is widespread from Imagi Mt near Iringa across to the Njombe and Mbeya highlands from 1 500 upwards.

This subspecies is also found in Kenya, elsewhere in Tanzania and southwards to Mozambique and Zimbabwe.

Wahlenbergia capitata (Bak.) Thulin [34B]

Sixteen of the 25 East African species of *Wahlenbergia* are found in our area, and, of these, *W.capitata* is one of the commonest being first collected on the Nyika Plateau in neighbouring Malawi.

It is an annual or more commonly short-lived perennial herb up to 1 m tall growing from a taproot. The leaves are 1.5 to about 6 cm long and the blue, white or mauve flowers 5 to 6.5 mm long. The latter are scarcely bell-shaped for they are lobed almost to the base and on the outside they are hairy on the mid-veins of each lobe.

Plate 34. A, *Wahlenbergia capillacea;* B, *W.capitata;* C, *W.subaphylla*

In the Southern Highlands it may be found from 900 m growing in savanna woodland or bushland right up to the mountain grasslands at 2 400 m. It is also not infrequent in areas of old cultivation. A typical locality is by Lake Ngwazi at Mufindi where it flowers from March until May.

W.capitata is found elsewhere in western and southern Tanzania and also in Uganda and Zaire southwards to Zimbabwe and Mozambique.

Wahlenbergia subaphylla (Bak.) Thulin [34C]

On the way from Chimala to Matamba in the Uanji country, the road crosses the Chimala River at about 2 000 m after a hectic climb up the escarpment above Chimala involving some 55 hairpin bends! On the rocky ground leading down to the river, *W.subaphylla,* yet another species of this rather bewildering and ubiquitous genus can be seen in flower in April.

An erect perennial herb up to 50 cm tall, *W.subaphylla* has leaves 1.5 to about 7 mm long. The lax inflorescences are much branched and bear small white or bluish flowers, 5.5 to 9 mm long, the corolla being deeply divided into narrow lobes.

Only subspecies *thesioides* Thulin of *W.subaphylla* grows in our area, the typical subspecies being found only in Madagascar and subsp. *scoparia* (Wild) Thulin being confined to the border mountains of Zimbabwe and Mozambique. Subsp. *thesioides* is widespread in southern and western Tanzania and in neighbouring parts of Zaire, Malawi and Zambia.

Of the several other species of *Wahlenbergia* found in our area the virtually leafless *W.virgata* is probably the commonest being found everywhere above 1 100 m. *W.krebsii* is also a common and widespread species in the area.

Lobelia mildbraedii Engl. [35A]

The Ruwenzori, Mt Kenya and Kilimanjaro are noted for the many giant species of plants in genera such as *Senecio* and *Lobelia* which otherwise usually consist of quite small herbaceous species found at lower altitudes. However, some giant species are found in quite a few of the other mountain areas of East Africa although nowhere in quite the same density. One of the giant *Lobelia* species that reaches our area is the elegant 2 to 2.5 m tall *L.mildbraedii* also recorded from neighbouring northern Malawi and southern Zaire.

This outgrown herb has a hollow stem the lowermost 1 m of which is covered by numerous softly hairy leaves up to 30 cm long and 2 cm broad. The dense inflorescence consists of hundreds of flowers in a cylindric column up to 1.5 m long, the lowermost flowers opening first. These are dull purplish in colour and 4 to 4.5 cm long being subtended by a softly hairy bract half as long as the flower.

L.mildbraedii occurs only on the highest areas of the Kitulo Plateau above 2 200 m where it grows in swampy grassland in association with *Gunnera, Kniphofia, Geranium, Carduus* and *Satyrium* species. A typical locality is the

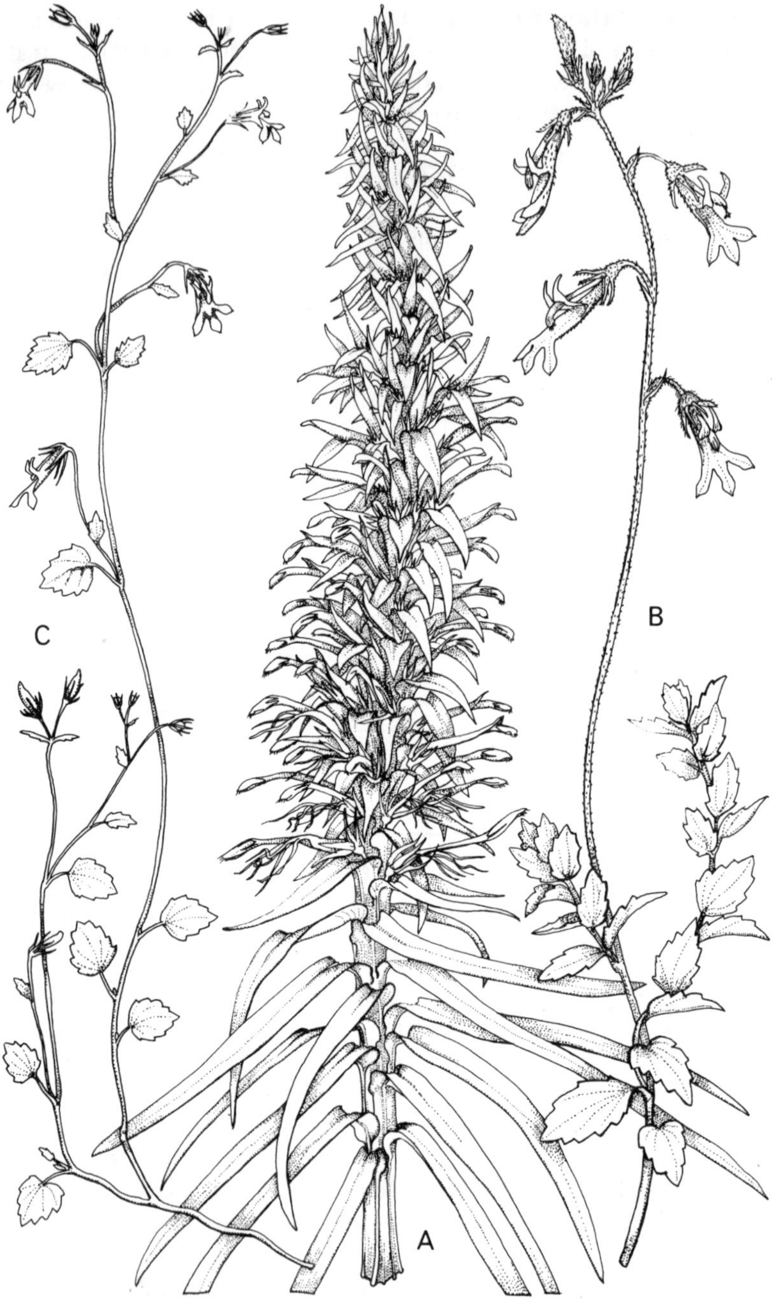

Plate 35. A, *Lobelia mildbraedii;* B, *L.ovina;* C, *L.trullifolia*

swamp 1 km SE of Kitulo Farm where many plants can be seen in a shallow valley leading down to the road.

In the montane forests of the central Southern Highlands an even taller species *L.gibberoa* is to be found. This is called 'malonge' in Kinga, 'piriti' in Malila, 'izyunzyu' in Safwa, 'ibuto' in Nyakyusa and 'ndelelengu' in Ndali. It flowers only after several years of growth. Numerous examples are to be seen on the roadside between Igoma and Kitulo.

Lobelia ovina E.Wimm. [35B]

A branching, hairy subshrub, 18 to 60 cm tall, growing from a rootstock 5 mm in diameter. The leaves borne at the base of the stem reach 1 to 1.3 cm long and 0.7 cm in diameter. Each inflorescence consists of 2 to 5 pinky mauve flowers, 1.9 to 2.4 cm long, borne on long stalks.

Mary Richards collected it in January 1957 on Mtorwi Mt the highest point of the Kitulo Plateau and in the Southern Highlands. She found it growing there in association with the beautiful white-flowered orchids *Satyrium buchananii* and *Habenaria macrura,* the poker-like yellow and orange-flowered *Disa ochrostachya* and various other colourful flowers. Elsewhere *L.ovina* has been recorded in flower between September and January at altitudes as low as 1 900 m.

Several other *Lobelia* species occur in the Southern Highlands mostly at lower levels. One other high-altitude rarity is *L.adnexa* recently collected for the first time in Tanzania below Irambo in the Poroto Mts in May.

Lobelia trullifolia Hemsl. [35C]

A weak, erect, annual herb with branching stems up to 35 cm long. Its small 1 cm long leaves are borne on slender 6 mm long stalks. The flowers are 1 cm long and are predominantly pale blue with a yellow base to the lower petal, which has a central white patch and two dark blue spots between the lobes. The stigma in the centre of the flower is dark bluish green.

L.trullifolia is often found growing with the pink-flowered *L.goetzei.* A typical locality is on the eastern side of the Kitulo Plateau where the great escarpment slope falls to the Mbarali River at 1 500 m. The edge of the plateau consists of an area of Buanji (or Bukoban) sandstones. It can be reached from below from Wangama village via the Fulaningi Ridge. This rocky ridge is alive with flowers in July with *Gnidia, Gladiolus laxiflorus, Moraea thomsonii, Kalanchoe lanceolata, Aloe nuttii* and *Wahlenbergia* and *Selago* species all in full bloom. Right at the top, on the edge of the Plateau, dense masses of *L. trullifolia* and *L.goetzei* can be found growing together under rocks. The view from the top is magnificent and well worth the long climb.

We recently also saw it in the magnificent setting of Kipipa Mt near Kisinga in the Ukinga Highlands. In early May at the end of the rains the whole of its grassy slopes were covered with tens of thousands of this tiny blue species growing in dense clumps.

Plate 36. A, *Kohautia coccinea;* B, *Pentas lanceolata;* C, *Spermacoce dibrachiata* (C1, flower)

RUBIACEAE
THE COFFEE FAMILY

Plates 36 and 37

A very large family of herbs, shrubs or trees. The shrubby species are particularly prominent as the understorey in the forests of tropical Africa. Leaves opposite, simple and entire and always stipulate. Flowers are borne in panicles, cymes or congested heads or sometimes solitary. Flowers regular, bisexual or less frequently unisexual, with three to eight free sepals and three to 11 petals and stamens. The last are borne on the corolla tube whose mouth is frequently filled with flattened, ribbon-like hairs. The ovary is inferior (except in *Gaertnera*) and the fruit is a capsule, drupe, berry or schizocarp.

The best-known members of this family are coffee (*Coffea* species) and quinine (*Cinchona* species) both frequently cultivated in Tanzania.

Kohautia coccinea Royle [36A]

This brilliantly coloured plant is one of a select band of species which occur in both Africa and on the Asiatic mainland, in this case in the Himalayas of India.

It is an erect annual herb with simple or weakly branching stems up to 70 cm tall. The linear leaves may vary from 1 to 7.7 cm long and up to 8 mm wide. However, its unmistakable flowers catch most plant-hunters' eyes sooner rather than later for these are scarlet or crimson or less commonly pink, the corolla tube being 2 to 6 mm long and the lobes up to 7.5 mm long.

It is widespread in the Southern Highlands above 1 000 m altitude flowering from April until June and may be easily seen at Sao Hill and a few km west of Mbeya in grassland and bush on the roadsides. Elsewhere it is widespread throughout tropical Africa up to about 2 400 m altitude from Ethiopia south to Mozambique and Zimbabwe.

Pentas lanceolata (Forssk.) Deflers [36B]

The genus *Pentas* includes some of the showiest flowers found in the Southern Highlands. About ten species of the genus are found here and most are recognisable by their large white to pinky mauve or violet trumpet-shaped flowers. *Pentas lanceolata,* figured here, is characteristic of the genus. It is a very variable hairy herb or subshrub with erect branches, up to 1.3 m tall, and paired leaves from 3 to 13 cm long and 1 to 6 cm broad. The flowers, borne in a terminal cluster, are pinky mauve and large, up to 4 cm long and 0.6 to 2 cm across. The throat of the flower is hairy.

P.lanceolata is an extremely variable species which has been divided into two subspecies and several varieties throughout its distribution which ranges

Plate 37. A, *Fadogia odorata;* B, *Otiophora caerulea;* C, *Rubia cordifolia*

from the Yemen and Ethiopia southwards throughout East Africa. Only subsp. *quartiniana* (K.Krause) Verdcourt have been recorded from our area where it is a common plant of roadsides, abandoned cultivation, secondary bushland and forest clearings up to about 2 000 m.

Perhaps the showiest species in our area is the *P.decora* S.Moore, a beautiful and common grassland species with whorled leaves and large sweetly scented white trumpet flowers reaching 13 cm long. It is frequent on the Mbeya and Poroto ranges and Kitulo Plateau. The Safwa people of the Porotos call it 'fwalo', the Kinga people further east call it 'manywagu' and the Wanyiha, 'impangwe'. It is the common Christmas-time flower of these areas.

Spermacoce dibrachiata Oliv. (syn. *Borreria dibrachiata* (Oliv.) K.Schum.) [36C]

Over 30 species of *Spermacoce* have been found in East Africa and, of these, 12 have been recorded in our area. Some of these are widespread and some have even become weeds of cultivation. *S.dibrachiata* is one of the most widespread species and brightens up the hillsides and woodlands with its blue or violet-blue flowers long after everything else has dried up. It is an annual or possibly a biennial herb, to 75 cm tall, with single or sparsely branching stems sparsely to densely covered with spreading hairs. The hairy leaves are 4.5 to 12 cm long and up to 2.8 cm broad. The flower heads reach 2.5 cm across and the individual flowers are 1 to 2.5 cm long.

S.dibrachiata is found throughout Tanzania and from Zaire and Rwanda south to Zimbabwe and Angola at altitudes up to 2 250 m growing in a variety of habitats from woodland to grassland and even in cultivated places. We have found it at Milo in Upangwa flowering in June in deep grass at the edge of a thicket growing in red sandy soil. The Wanyiha call it 'ikuwe' and 'ilungalunga'.

Fadogia odorata K.Krause [37A]

In the upland grasslands near Sheya village in the western Bundali Hills at about 2 100 m the vegetation is dead and brown by August some two months into the dry season. However, the ground is brightened by the flowers of *Leonotis, Hypericum* and *Dissotis* species whilst the white flowers of *Fadogia odorata* pepper the ground over large areas.

This pretty species sends up its shoots to 30 cm tall from a woody base. The leaves borne in whorls of three or four reach 5 cm long and 1 to 2.5 cm broad. The inflorescences are axillary on pedicels up to 4 cm long and bear white flowers about 1 cm across with conspicuous brownish anthers.

F.odorata is widespread in the Southern Highlands from Mufindi westwards and the specimen illustrated here was collected by Mary Richards between Sumbawanga and Chala where she found it growing in a carpet of flowers brought on by the first rains of the season in October. Another species in the same family also flowering here was the pale blue-flowered *Pentanisia schweinfurthii.*

Otiophora caerulea (Hiern) Bullock [37B]

The genus *Otiophora* with about nine species in East Africa includes annual and perennial herbs and subshrubs with erect, straggling or creeping stems. All of these species occur in our area.

O.caerulea figured here is a perennial herb with three to five erect shoots, more or less covered with bristly hairs and up to 45 cm tall, arising from a divided woody rootstock. The leaves are commonly paired but are variable in shape and range from 1.7 to 5 cm long and 0.2 to 1.1 cm broad. The inflorescences are compact heads and terminate the main and side branches. These bear blue to mauve or mauvy pink flowers up to 1.2 cm long.

This pretty species is to be looked for in the woodlands and montane grasslands above 1 500 m throughout the area during the rainy season from November up until July. Typical localities are at Kibengu, south of Dabaga, where it grows in grassland over dark red soil at 2 150 m and on the Mbeya to Chunya road where it grows at 2 100 m in deep loamy soil on the steep hillsides of Chunya escarpment.

Rubia cordifolia L. [37C]

A widespread weedy climbing or creeping plant with a slender brittle stem and readily recognised by its prickly recurved hairs. The long-stalked, 5-veined leaves, borne in whorls at each node, are about 4.5 cm long and 3 cm broad. Greenish-yellow flowers about 4 to 6 mm in diameter, are borne in short axillary cymes.

Plants from East and Southern Africa have been assigned to subsp. *conotricha* (Gandoger) Verdc. identified by its leaves which are borne in whorls of four and are hairy below and by its fruits, the cocci of which do not exceed 3 mm in diameter.

It is common throughout the area from 1 100 to about 2 600 m being found on forest margins, in clearings in the forest, in scrub and in grassland.

The Wamalila people in the highlands south of Mbeya call this herb 'ibambula' and the Nyiha to the west 'ivambula'.

VALERIANACEAE
THE VALERIAN FAMILY

Plate 38A

Unpleasantly scented herbs or shrubs with opposite, often divided leaves which usually have clasping bases. Inflorescences cymose, rather densely many-flowered, the flowers bisexual and irregular. Calyx five-lobed, obscure until the fruit develops when it unrolls into a 'parachute' of plumose segments. Corolla-tube long, usually five-lobed, the one to four stamens being attached to the upper part of the tube. Ovary inferior, the fruit a dry indehiscent cypsela.

Valeriana capensis Thunb. [38A]
A perennial offensive-smelling herb, up to 180 cm tall but commonly less than 100 cm, with brittle stems and a stout rhizome. The leaves are 8 to 20 cm long and are borne on a stalk up to 35 cm long. The flowers are pinkish-mauve in bud but open white or pinkish-white. The corolla tube is 2 to 5.5 mm long.

It is the only species of *Valeriana* found in the area, where it is widespread. It grows on both Mt Rungwe and on the Kitulo Plateau, from about 2 200 m upwards, in open grassland and in marshy areas. It can also be seen at Mdapo near Njombe. It flowers in our area from December until about May.

As its name suggests this species was originally described from the Cape but its range extends northwards to Kenya.

Plate 38. A, *Valeriana capensis;* B, *Cephalaria goetzei;* C, *C.pungens*

DIPSACACEAE
THE SCABIOUS FAMILY

Plates 38B and C

Herbs or subshrubs with entire or lobed leaves lacking stipules. Flowers in dense heads subtended by involucral bracts, the marginal flowers often longer. They are bisexual or female and irregular, each with a basal epicalyx of fused small bracts. Calyx cup-shaped, four- or five-lobed as is the corolla. Stamens two or four, growing on and alternate with the corolla lobes. Ovary inferior and developing into a dry indehiscent fruit.

Some species bear a superficial resemblance to the Compositae but are distinguished by their protruding free anthers and fruit.

Cephalaria goetzei Engl. [38B]

A perennial herb, to 60 cm tall, with erect glabrous or sparsely hairy stems. The basal leaves are variable in shape, lobing, size and pubescence but the upper, however, are generally all alike and about 4.5 cm long. Flower-heads subglobose, 2 to 3 cm in diameter; the silvery pubescent bracts with dark tips subtending white flowers.

It grows in upland grassland and marsh from 1 500 to 2 400 m altitude. It is not infrequent on the Mbeya Range especially on the Chunya escarpment and is also common around Mbozi. Elsewhere, it is found in parts of western and southern Tanzania and also in neighbouring Malawi.

It is closely allied to *C.integrifolia* Napper which is also found in the Southern Highlands from Dabaga westwards. The latter differs though in having villously hairy long-toothed young fruits.

Cephalaria pungens Szabó [38C]

A perennial herb, 50 to 100 cm tall, with erect unbranched glabrous or sparsely pubescent stems. The glabrous or pubescent leaves are variable in shape and size but the lower ones can reach 35 cm long. The flower-heads are subglobose, 2 to 4 cm in diameter, with pubescent bracts and white flowers, the marginal ones usually larger than the inner ones. The sparsely hairy young fruits have four short terminal lobes.

It is a far more wide-ranging species than the previous one being found as far north as Kenya and southwards to Lesotho. Its habitat is upland grassland, often in marshy places, and also glades in montane forest above 1 800 m. In our area it can be seen in flower from March until May on the Kitulo Plateau at between 2 600 and 2 900 m.

Plate 39. A, *Ageratinastrum polyphyllum;* B, *Artemisia afra;* C, *Aster tansaniensis;* D, *Athrixia rosmarinifolia*

COMPOSITAE
THE SUNFLOWER FAMILY

Plates 39 to 45

Probably the largest family of flowering plants with possibly 25 000 species. Perennial rhizomatous herbs, subshrubs, shrubs or rarely trees. Leaves alternate or opposite, lacking stipules. Flowers borne in a head called a capitulum made up of a few to many small individual flowers surrounded by an involucre of protective bracts. In many capitula the inner flowers (called disc florets) are fertile and the outer ones (ray florets) sterile with one petal well-developed so that the whole capitulum resembles a single flower. In fertile flowers the calyx is made up of hairs, scales or bristles which help in the dispersal of the fruit. Corolla of fused petals bearing five united stamens; ovary inferior. Fruits one-seeded, dry and indehiscent.

Economically important species include lettuce (*Lactuca sativa*), chichory (*Cichorium intybus*), globe artichoke (*Cynara scolymus*), Jerusalem artichoke (*Helianthus tuberosus*) and sunflower (*H.annuus*). Many species are extremely ornamental and have been widely introduced into cultivation.

Ageratinastrum polyphyllum (Bak.) Mattf. [39A]

This is a shrub, 60 to 100 cm tall, with erect brown hairy stems arising from a woody rootstock. The leaves, crowded towards the tops of the stems, are 1.7 to 4 cm long and 6 to 9 mm broad, green above but greyish on their undersurface. At the tops of the stems are numerous small mauve or purple flower-heads, each 1 to 1.1 cm in diameter.

Mary Richards collected it in March at Madunda in Upangwa and again in April on Kito Mt in Ufipa growing on grassy slopes at 1 800 m. We have found it on Mbogo Mt in Umalila at 2 300 m. The summit moors of Mbogo are dotted with its pretty flowers from April until August. This mountain, named after the buffalo which was formerly common here, dominates the landscape looking south from Mbeya. It can be climbed in two or three hours from the village of Ilembo in Umalila. The summit view is magnificent extending to the Ufipa Highlands, Bundali Hills, Mbeya Range and the Rukwa, Mt Rungwe, the Livingstone Mts and a wide sweep of northern Malawi.

A.polyphyllum is also widespread in Malawi and Zambia from 1 500 m upwards. Another species of the genus found in the Southern Highlands is *A.goetzeanum* which is also widespread.

Artemisia afra Jacq. [39B]

This plant, a species of wormwood, a genus familiar in Europe where several species are renowned for their medicinal properties, is likewise noted for its medicinal and magical uses. Consequently, every tribe has a name for it – the Safwa, the Kinga and the Malila calling it 'inyaga', 'lunyaga' and 'ilumbati' respectively.

It is an erect shrub, to 2 m tall, with aromatic silvery grey leaves up to 6 cm long and nodding ochre-yellow or pale yellow flower-heads, 2 to 4 mm long, borne at the top of the stems in elongated inflorescences.

In places in the Southern Highlands it can be a dominant plant and we have found it so in some parts of the Kitulo Plateau at about 2 400 m where it flowers in May.

Aster tansaniensis Lippert [39C]

This is an erect herb, 13 to 50 cm tall, often found growing in clumps. It has hispid stems and leaves, the latter up to 5.5 cm long and 7 to 9 mm broad also having spiny margins. The solitary flower-heads, about 4.5 cm across, have a central yellow disc surrounded by lavender-pink, pale mauve or white ray florets. The involucral bracts which enclose the base of the flower-head are green.

In October and November, the moorlands of the Kitulo Plateau are decorated with millions of *A. tansaniensis* in full flower. It is widespread elsewhere in the Southern Highlands, flowering with the start of the rains from October onwards. It commonly grows in association with several other composites such as *Vernonia smithiana, Helichrysum latifolium, Hirpicium angustifolium* and *Gerbera viridifolia* and other early flowers such as *Gnidia involucrata, Monadenium orobanchoides, Vigna pygmaea, Polygala ohlendorfiana* and *Euphorbia cyparissoides.* These flushes of early flowering species, often appear on burnt ground after the early rains and are a marvellous sight nowhere better seen than on the Kitulo Plateau.

Elsewhere in Africa, *A. tansaniensis* is found in Malawi, Zambia and Southern Zaire.

Athrixia rosmarinifolia (Sch.-Bip. ex Walp.) Oliv. & Hiern [39D]

This is an erect much branched plant with brown woody stems, 45 to 120 cm tall, and grey-green foliage. The lower leaves reach 6 cm long and 5 mm broad but the upper leaves are much smaller and narrower, only 2 to 3 cm long and 1 or 2 mm broad. The flower-heads, borne at the ends of the branches, are up to 1.4 cm in diameter enclosed by an involucre of grey bracts. Around their yellow or orange central disc florets are lilac, pink, purple or rarely yellow ray florets.

It is widespread in the mountains of the area where it often grows in association with other Compositae such as *Stoebe kilimandscharica* and *Artemisia afra* and the shrubby legume *Kotschya recurvifolia.* A typical locality is on the Safwa Hills where it occurs on the higher slopes above 2 000 m flowering from

July until October. There the local people call it 'matungampanzi'. Elsewhere in our area it is found from Mufindi where the Hehe call it 'malolo' across to Ufipa. It is also widespread in south-central Africa south to Zimbabwe and elsewhere in East Africa north to Ethiopia.

Berkheya echinacea (Harv.) O.Hoffm. ex Burtt-Davy [40A]
This plant covered by soft bristle-like rusty hairs on all its parts is rather like a thistle. It is a large herb with erect single stems 60 to 100 cm long, the lower leaves up to 13 cm long and 3 cm broad below but the upper ones smaller. Flower-heads yellow, 5 to 6.5 cm across when fully open, with an involucre of softly hairy bracts about 2 cm long.

Plants from Tanzania and south-central Africa belong to subsp. *polyacantha* (Baker) Roessler which differs from the typical subspecies found in South Africa in its larger size and much longer and broader leaves.

It is widespread in the Southern Highlands, in western Tanzania and in neighbouring Malawi and Zambia. On the Malonje Plateau, Ufipa it grows in grassland on fire-swept hillsides at between 2 200 and 2 700 m. It can also be readily seen on the Mbeya Range, on Mt Rungwe, around Njombe and in west Mufindi always growing in grassland and mostly above 1 650 m. It flowers from March onwards in our area.

Bidens schimperi Sch.-Bip. ex Walp. [40B]
Despite its weedy propensities this is nevertheless an attractive species. It is usually a straggling to erect annual herb, 50 to 120 cm tall, with reddish commonly laxly branched stems somewhat four-angled in section. It has pinnatifid leaves, up to 22 cm long, borne on a hairy stalk. At the ends of the stems are the flower-heads, 3 to 5 cm in diameter, with a brownish yellow central disc surrounded by bright yellow ray florets. The involucral bracts are pale green and hairy.

This species is found throughout our area as a weed of cultivation, on roadsides and in open woodland. At Mbozi, west of Mbeya, it is in flower in May growing along the roadsides by open swampy areas at 1 500 m. It is particularly common around Sao Hill and Mufindi to the southwest of Iringa and can easily be seen along the main road to Mbeya between Mtanana and Malangali from March until August. It grows in more natural surroundings on the Ndala Pass on the road leading from Mwakaleli in the Unyakyusa Highlands northwards to the Porotos. There, it flowers in July on the margins of forest patches on the lower slopes of Mt Rungwe. The Wasafwa call it 'imbotila' and the Wanyiha 'mpangwe'. It is widespread throughout south-central Africa.

B.schimperi is only one of several species of the genus found in our area. Others include *B.dolosa, B.pilosa, B.ugandensis, B.magnifolia* and *B.lineariloba. B.pilosa* is the ubiquitous 'black-jack' weed of gardens and disturbed ground. In contrast, *B.pinnatipartita,* growing to 3 m tall on the edges of forests of the

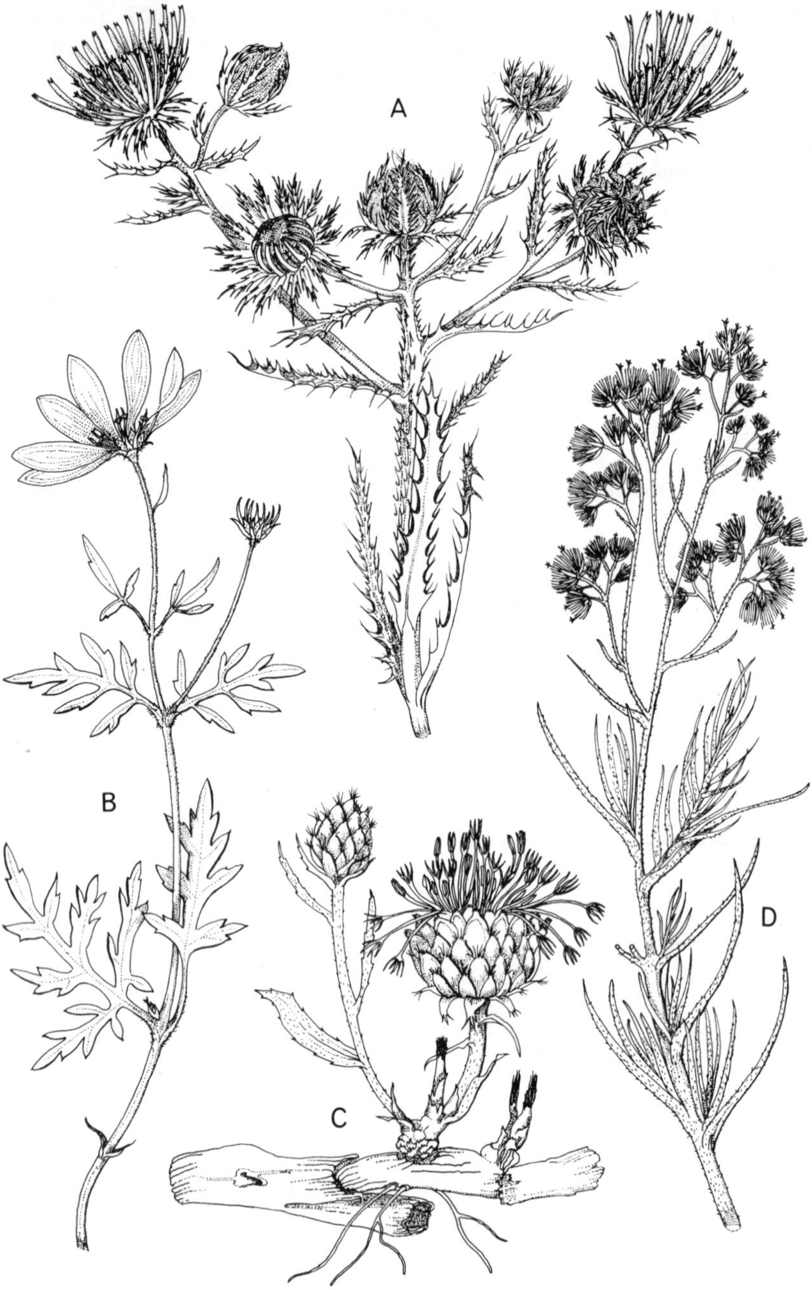

Plate 40. A, *Berkheya echinacea;* B, *Bidens schimperi;* C, *Centaurea praecox;* D, *Nidorella welwitschii*

area is a wonderful sight in May. In the heath forests on the north slopes of Rungwe Mt, it grows in association with the white-flowered tree heath *Erica arborea.*

Centaurea praecox Oliv. & Hiern [40C]

C.praecox, a widespread species in the Southern Highlands, is a hairy perennial with 2 to 3 cm long stems or branches arising from a woody rootstock. The few oblanceolate leaves have finely saw-toothed edges and reach 3 cm long. The flower-heads, borne singly, are about 3 cm long and wide. At the base of these are many spine-tipped bracts which enclose the bases of numerous purple florets, each up to 2 cm long.

Mary Richards recorded this species in flower in early September from the Muse Gap in Ufipa and also in the Nsangu Forest. A typical locality is at Ilindi on the north side of the Umalila range at 1 650 m. In late August, when the grass has been burnt and the first rain showers have begun, the open ground is a veritable garden of flowers including this *Centaurea,* the beautiful blue-flowered *Trichodesma hockii* and the white-flowered *Senecio pachyrhizus.*

Nidorella welwitschii S.Moore [40D]

This common species is rather variable in stature, its pubescent stems, ranging from 40 to 200 cm tall, often branching above to form a small bush. It has minutely pubescent leaves about 5 cm long and 6 to 9 mm broad and yellow flower-heads about 9 mm in diameter.

It is widespread in the Southern Highlands and can be seen in flower from October until February. At both Sao Hill and Mufindi, it grows in meadows and on the edges of marshland at about 2 000 m. Further west, it can be seen on the Kitulo Plateau up to 2 700 m and in the grassland on the slopes of Ngosi Mt in the Porotos at 2 100 m. Mary Richards found it in January 1957 growing amongst tall grasses on the margins of the Lihogosa swamp near Njombe. This area is still poorly known and would repay a systematic examination throughout the wet season.

This species is widespread elsewhere in East and south-central Africa. Other species of *Nidorella* found in the Southern Highlands are *N.spartioides* commonly seen on the Mbeya Range and also at Mufindi, *N.auriculata* in the Kipengere Mts, *N.stricta* with pale green flowers in the Mbisi forest near Sumbawanga, and the orange-flowered *N.microcephala,* a low-altitude species in the north Rukwa area and Madibira Mission.

Cineraria grandiflora Vatke [41A]

This ubiquitous species is variable in habit and can be found as an erect bush or as a scandent or trailing plant. Its purplish hairy stems range from 35 cm to 1.2 m tall. The leaves, commonly 4 to 5 cm long and 3 to 4-5 cm broad, borne on a hairy stalk about 3 cm long, are hairy beneath particularly along the veins.

Plate 41. A, *Cineraria grandiflora;* B, *Gerbera viridiflora;* C, *Haplocarpha thunbergii*

The yellow flower-heads, commonly about 1 to 1.5 cm in diameter, decorate the countryside in the dry season from July until September and are much sought after by bees. In places it colours the roadsides and bush above about 1 800 m altitude.

It is common around the upper fishing camp on Mt Rungwe and throughout the Poroto and Mbeya ranges, up to about 2 550 m, growing in grassland, on the margin of dense forest and even as a weed of neglected cultivation. The Wamalila call it 'msinde'.

Elsewhere in East Africa it is widespread and common as far north as the Sudan and Ethiopia.

Gerbera viridiflora (DC.) Sch.-Bip. [41B]

This plant, better known under the later name *G.abyssinica* Sch.-Bip. ex A. Rich., has a basal rosette of leaves, up to 19 cm long and 6.5 cm broad, which are silky hairy at the base of their long stalks. The solitary flower-head, varying from 3.5 to 5 cm across and borne on the apex of a hairy stalk 9 to 26 cm long, has a white centre and purple- or pink-tinged ray florets. It is widespread in the Southern Highlands area from Dabaga in the east across to Rungwe and Mbogo Mts in the west and also north to Sumbawanga. We have found it in flower on the grassy SE slopes of Mbogo Mt in Umalila at between 2 200 and 2 400 m in November. It can also be seen on Rungwe Mt around the top fishing camp at 2 350 m in September and Mary Richards collected it at Kikondo one of her favourite camping spots on the edge of the Kitulo Plateau.

Elsewhere it is widespread in East Africa as far north as the Sudan and Ethiopia.

Haplocarpha thunbergii Less. [41C]

This plant, perhaps better known in our area under the later name *H.scaposa* Harvey, is herbaceous, 15 to 50 cm tall, with dark green leaves, up to 27 cm long and 6 cm broad, which are white-felted on their undersurface. The yellow flower-head, 4.5 to 5.2 cm in diameter, is carried on a hairy stalk.

It is a widespread and common grassland species in the Southern Highlands above 2 000 m and one of the many flowers which spring up after the grassland has been burnt, flowering from November onwards. Typical habitats are on the Kitulo Plateau up to about 2 800 m, in the Poroto Mts where it is often found as a weed of *Pyrethrum* cultivation and around Lake Ngwazi and the Sao Hill marshes further east.

It is widely distributed in south-central and South Africa as far as Cape Province.

Helichrysum herbaceum (Andr.) Sweet [42A]

This is another widespread species of the Southern Highlands area much better known under the later name *H.squamosum* Thunb.

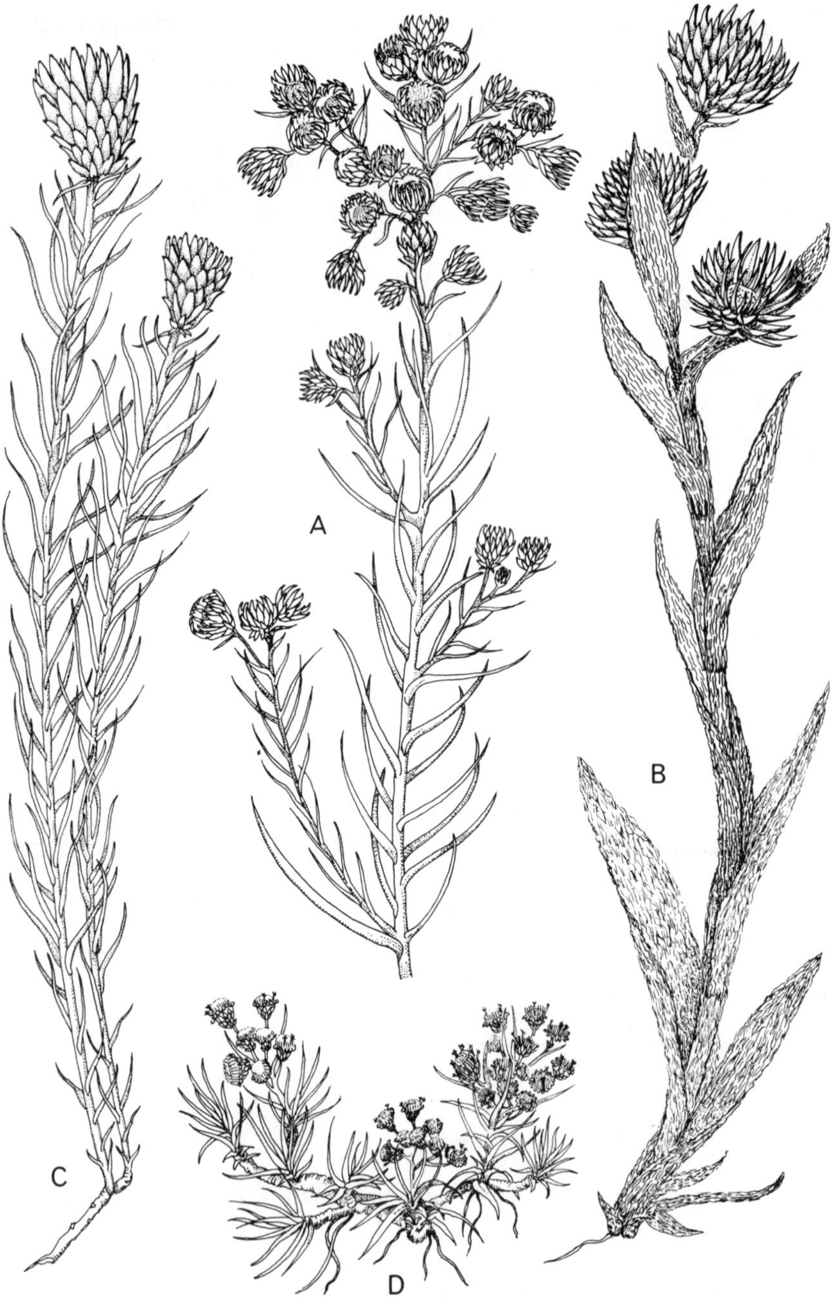

Plate 42. A, *Helichrysum herbaceum;* B, *H.nitens;* C, *H.kirkii;* D, *H.tillandsifolium*

Its ascending stems, 15 to 40 cm tall, are grey-felted all over. The linear leaves, likewise grey-felted, are 1 to 2.6 cm long and 1 to 2 mm broad. The stems terminate in solitary large flower-heads up to 2.6 cm in diameter; the florets yellow or orange, surrounded by pale lemon bracts tinged with brown.

It can be seen in its full glory in the Sao Hill grasslands in January and February at 1 900 m altitude. It is found throughout the Kipengere Mts and Kitulo Plateau and also on the Mbeya Range in similar grassland areas until May.

H.herbaceum is found also in the Uluguru Mts of Tanzania and is widespread in south-central and South Africa.

Helichrysum kirkii Oliv. & Hiern [42C]

Helichrysum, the species of which are popularly called 'everlasting flowers', is one of the largest genera of the Compositae in our area and certainly one of the most attractive.

H.kirkii, one of the commoner species, is a small perennial herb or shrub with erect clustered stems, about 40 to 100 cm tall, arising from a woody rootstock. The sessile hairy leaves are 2 to 3 cm long and 2 to 3 mm broad. Bright yellow flower-heads, 1 to 1.8 cm in diameter, cluster at the top of the stems.

It is common in upland grasslands between 1 800 and 2 100 m along the road from Mbeya to Chunya, along which there are marvellous views over the Rukwa valley on the left and the Usangu flats on the right. *H.kirkii* grows here in association with the purple-flowered *Ageratinastrum polyphyllum,* the drooping pink flowers of *Dierama pendulum* and the small shrubby *Protea heckmanniana.* It is also common on the Kitulo Plateau and in the Poroto Mts at about 2 300 to 2 400 m and can be seen in flower from May until July.

It is common elsewhere in the mountains of east and central Tanzania and throughout south-central Africa.

Helichrysum nitens Oliv. & Hiern [42B]

This is an unmistakeable species readily recognised by its densely white- or grey-felted alternate leaves, up to 14 cm long and 0.7 to 2 cm broad, and its stems, about 20 to 40 cm tall, are also densely hairy. The yellow flower-heads, arising near the top of the stems, are 2 to 3 cm in diameter and surrounded by straw-coloured involucral bracts.

It flowers in the rainy season on the higher ground of the Southern Highlands, mostly above 2 100 m, and on Mbeya Peak can be seen in flower in April in the grassland at 2 750 m. Mary Richards found it in flower in Ufipa on Nsanga Mt on the Malonje Plateau at 2 100 m in March.

H.nitens is recorded also from the mountains of north-east Tanzania and from similar areas in south-central Africa.

Helichrysum tillandsifolium O.Hoffm. [42D]

This is another unmistakeable species readily identified by its habit of forming

Plate 43. A, *Helichrysum sulphureo-fuscum;* B, *H.wittei;* C, *Hirpicium angustifolium*

low cushions 3 to 9 cm high. It is a branching subshrub growing from a long woody rootstock. Its leaves, which are white-felted below, are 2 to 4 cm long and 2 to 5 mm broad. Some of these droop but the upper ones are held erect encircling the flower stems. The tiny yellow flower-heads, 3 to 5 mm in diameter, have papery involucral bracts and are borne in an inflorescence 2.5 to 4 cm wide.

It is widespread in the central higher areas of the Southern Highlands, for example, on the north side of the Kitulo Plateau and on the Mbeya Range up to the summits. It can be seen in flower from September until March in these areas.

Elsewhere, it has only been recorded from similar areas of northern Malawi.

Helichrysum sulphureo-fuscum Bak. [43A]

This perennial subshrub has numerous decumbent and ascending stems, 20 to 50 cm long, arising from a woody rootstock. Its spreading leaves, clustered at the base and less frequent along the stem, are 2 to 3 cm long and 2 to 3 mm broad, pale green above and whitish beneath. Each stem ends in a solitary bright yellow to orange-yellow flower-head, 1.5 to 2 cm in diameter; the lemon-yellow to pale pink bracts distinctively tipped dark grey or black.

We have found it in flower in June growing in rocky ground on the northern side of the Bundali Hills in association with a pretty blue-flowered labiate *Leucas oligocephala.* These hills are a delight to the botanist with their short grassy slopes and plentiful flowers providing interest throughout the year.

This species can also be seen on Mbeya Peak at 2 520 m from April onwards, on Rungwe Mt around the upper fishing camp and on the Kitulo Plateau at about 2 400 m. Elsewhere it is also recorded from southern Tanzania in the mountains east of Lake Nyasa.

Helichrysum wittei Hutch. & B.L.Burtt [43B]

This is undoubtedly one of the most attractive of the everlasting flowers found in the area. It is a subshrub, 30 to 100 cm tall, with brown stems and leaves, 2 to 7 cm long, densely white-felted on their lower surface. Its silvery white flower-heads, not more than 1 cm in diameter, are pale yellow in the centre.

It can be seen in full flower in early July by the Ndumbi River at 2 800 m, one of the headwaters of the Great Ruaha flowing northwards off the Kitulo Plateau. Of particular interest here is the great forest of *Juniperus procera* which cloaks the northern slopes of Mtorwi Mt while trees of *Podocarpus* and *Salix subserrata* are also found here.

H. wittei has also been collected on Mbeya Peak in flower in May in a grassy area at 2 600 m. Elsewhere it is only known from western Tanzania and adjacent areas of Zaire.

Hirpicium angustifolium (O.Hoffm.) Roessl. [43C]

This small perennial herb seldom reaches 10 cm tall. At the base it produces a

Plate 44. A, *Felicia abyssinica;* B, *Lopholaena trianthemos;* C, *Notonia abyssinica;* D, *Osteospermum monocephalum*

number of slender flexuous spreading leaves up to 22 cm long but only 1 to 2 mm broad. The solitary yellow flower-head, about 3 cm in diameter, emerges on a short stalk from amongst the leaves. Its appearance is often triggered by burning.

Our first encounter with it was on Majembe Hill near Ibaba in the Bundali Hills in early August. The ground there is stony and we found it growing with the pretty pink-flowered orchid *Polystachya dendrobiiflora,* the red-flowered *Hibiscus rhodanthus, Moraea thomsonii* and *Gnidia kraussiana.* It is also common in December on the upper slopes of the Chimala escarpment at 2 100 m and in November again at Mufundi by Lake Ngwazi.

Felicia abyssinica Sch.-Bip. ex A.Rich. [44A]

Felicia is a large genus, of blue- or mauve-flowered daisies predominantly southern African in distribution, but a few species, such as *F.abyssinica,* are found in East Africa. This, a much-branched plant, 15 to 30 cm tall, growing from a woody rootstock has linear hispid leaves to 2 cm long and 1.5 to 3 mm broad. The flowers, about 2 cm across, last but a day but are prettily coloured yellow in the centre with radiating pale mauve or pale blue ray florets.

It is widespread in the Southern Highlands from Imagi Mt, east of Iringa, across to the Mbeya area. In the latter, it has even been found on the now-disused golf course at Mbeya where it flowers from September until December. It commonly grows in grassland or on open dry stony soils from 1 400 m upwards. Several varieties have been described throughout its range which extends from Ethiopia south to Zimbabwe.

In our area this species has been better known under the later name *F.hyssopifolia* Berg.

Lopholaena trianthemos (O.Hoffm.) B.L.Burt [44B]

This is an erect shrubby plant often growing in clumps with stems reaching 1.2 mm tall. Its leaves are glabrous and are commonly 6 to 9 cm long and 2.5 to 3.6 cm broad. The white or cream-coloured flower-heads, 1.3 to 1.5 cm long are enclosed by an involucre of green bracts.

It is not uncommon around Mbeya from about 1 500 m upwards, often being found in old cultivations or in montane grassland. On the Chunya escarpment it flowers in August and September in areas of burnt grassland up to about 2 150 m and it is also common in the grassland around Mbozi. In Umalila, we have seen it in flower in early October growing with the beautiful shrub *Vernonia subuligera* O.Hoffm.

This species is also recorded from neighbouring Zambia. An allied species *L.whyteana* is common on the Kitulo Plateau while *L.dolichopappa* is also widespread in the Southern Highlands where the Kinga call it 'liposa' and use it as a cure for snake-bite.

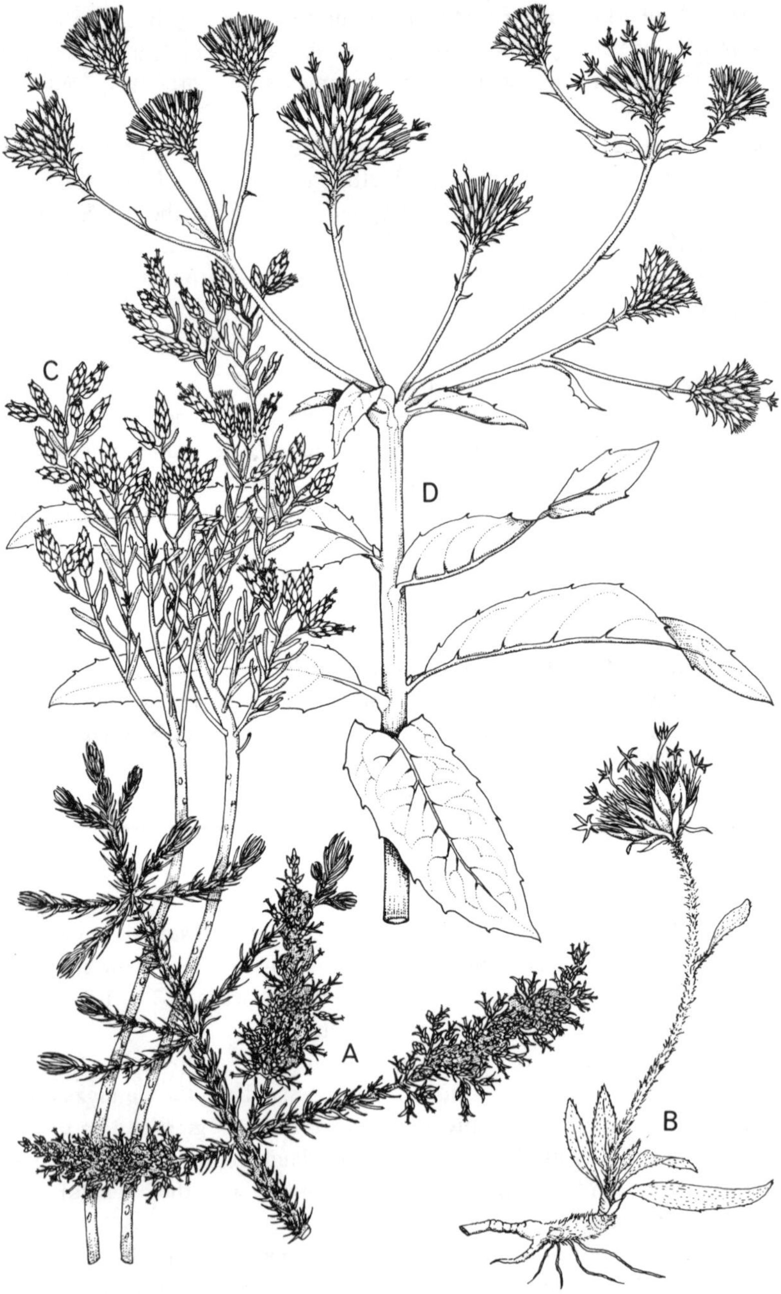

Plate 45. A, *Stoebe kilimandscharica;* B, *Vernonia nyassae;* C, *V.stenocephala;* D, *V.usafuensis*

Notonia abyssinica A.Rich. [44C]
This large glabrous succulent herb has coarse rather fleshy stems, up to 1.6 m tall, and rather gross obovate or oblanceolate leaves reaching 19 cm long. The scarlet flower-heads, about 4 cm in diameter, are loosely borne in terminal inflorescences.

It flowers at the end of the rains in May and June in the montane grasslands of the region from about 1 600 m upwards. The Wamalila people call it 'itivwa' and use the leaves to draw out blood.

Osteospermum monocephalum (Oliv. & Hiern) T.Norl. [44D]
This is an herbaceous plant with stiff brown stems, 10 to 20 cm tall, arising from a woody rhizome. The leaves, green, edged with brown, are shortly hairy, about 2.2 cm long. Its solitary yellow flower-heads, about 2 cm in diameter, are subtended by green involucral bracts. The flower stalks elongate to about 25 cm after fertilisation and the winged fruits are greenish conspicuously marked with red.

It is widespread throughout the Southern Highlands and Mary Richards found it on Kito Mt in Ufipa and at Kikondo on the Kitulo Plateau in flower towards the end of the dry season in September. In late October on Irundi Hill on the western side of Mufundi, this species is one of the most conspicuous plants. The swampy grasslands around Irundi drain into the headwaters of the Little Ruaha River. It flowers in burnt-over grassland there with *Scutellaria paucifolia, Euphorbia zambesiana,* the abundant yellow-flowered *Berkheya zeyheri* and *Launaea nana.*

Stoebe kilimandscharica O.Hoffm. [45A]
This curious wiry heath-like shrub, up to 4 m tall, has grey-green much-branched stems and pale silvery-green foliage. The leaves are, in fact, densely pubescent and about 6 mm long. The small yellowish-brown flower-heads, 4 to 5 mm long, are borne in axillary clusters towards the ends of the branches.

S.kilimandscharica is common in the Poroto Mts and can be seen in flower in August in the grasslands on the lower slopes of Ngosi Mt at about 2 600 m. In the Njombe area, it can be seen flowering in May near Bulongwa where the Kinga people call it 'mapumba' or 'lugenge'. In Umalila, it grows, amongst other places, near the summit of Mbogo Mt and is called by the locals 'izaza'.

It is a widespread plant in southern Africa south to the Transvaal. O.Hoffman has assigned plants from our area to var. *densiflora* O.Hoffm.

Vernonia nyassae Oliv. [45B]
This belongs to a rather complex group of closely allied African species each with a rather limited distribution. A dwarf perennial herb rarely more than 25 cm tall, growing from a woody rootstock has leaves, 2 to 5 cm long and

about 1 cm broad, hairy below, borne in a basal rosette. The purple flowerhead, borne on a long hairy stalk, is 2 to 2.5 cm in diameter.

It is found growing on bare stony ground in short grassland from around Iringa across to Njombe and the Kitulo Plateau at above 2 000 m altitude. At Sao Hill, where it flowers in January and February, it can be found growing in short turf by streams and in wet meadows.

The closely related *V.acrocephala,* which differs in its hairier leaves and scape and serrate leaf-margins, is found growing in stony ground on the hills of the western Undali around Ibaba where it flowers in October.

Vernonia stenocephala Oliv. & Hiern [45C]

This is one of a whole complex of closely related species in one of the largest African genera of the family Compositae.

Its branching hairy stems are about 60 to 100 cm tall rising from a woody rootstock. The leaves, 2 to 5 cm long and up to 1 cm broad, are minutely hairy on their undersurface. Numerous elongate heads of dark mauve, blue or rarely white flowers, 1 to 1.5 cm long and 3 mm in diameter, terminate the stems; the flowers amost hidden by the downy grey-green involucral bracts.

We have found this species in flower in early August at 1 800 m on the hills above Mbeya town growing with *Athrixia rosmarinifolia,* also at Mufindi by Lake Ngwazi and by the road to Iringa from Sao Hill at 1 800 m.

Elsewhere it is known only in neighbouring Zambia.

About 25 species of *Vernonia* are known to occur in the Southern Highlands.

Vernonia usafuensis O.Hoffm. [45D]

This is a commoner species around Mbeya than *V.stenocephala* and rather more readily identified because of its distinctive prickly leaves. It is a tufted plant whose purplish stems, varying from 60 to 160 cm tall, branch only towards the apex. The densely arranged elliptic leaves, reaching 8 cm long and 3 cm broad, are pale green above and paler and scabrid below. Flower-heads, 2 to 2.5 cm in diameter and 2 cm long, have bright blue to violet florets emerging from the top of the grey-brown involucral bracts.

In May 1957, at the end of the rains, Mary Richards collected this attractive species on Mbeya Peak in association with the allied *V.porphyrolepis* and the yellow-flowered legumes *Crotalaria vartifolia* and *C.rogersii.*

V.usafuensis can also be seen in the Poroto Mts, on Rungwe Mt, in Umalila and the Bundali Hills but elsewhere it is recorded only from western Tanzania.

XYRIDACEAE
THE YELLOW-EYED GRASS FAMILY

Plate 46A

Annual or mostly perennial herbs with leaves linear and sheathing at the base normally produced in tufts at the tops of the rootstock. Flowers bisexual, slightly irregular, arranged in globose or cylindrical heads, each flower in the axil of a stiff or leathery bract. Of the three sepals, the inner one forms a hood over the petals while the outer two are small and keel-shaped. Petals fused into an apically three-lobed short or long tube. Fertile stamens three, borne on the corolla tube opposite the lobes. Ovary superior and corolla persistant on the capsule.

Xyris nivea Welw. ex Rendle (syn. *X.obscura* N.E.Br. and *X.brunnea* Nilss.) [46A]

A bulbous plant growing in clumps with erect slender leaves 20 to 40 cm long and 0.5 to 3 mm broad. Inflorescence erect, with heads 6 to 15 mm long and 6 to 7 mm in diameter, blackish bracts and yellow flowers.

Found in damp areas throughout the region. We have found it flowering in September in a swamp at 1 850 m between Kigogo and Ndaga in Mufindi and also in January on Rungwe Mt in flushes below rocks at 2 200 m altitude. It is widespread elsewhere in Africa from Cameroun and Kenya to south-central Africa.

Plate 46. A, *Xyris nivea;* B, *Hypoxis angustifolia;* C, *H.urceolata*

HYPOXIDACEAE
THE HYPOXIS FAMILY

Plates 46B and C

Herbs with tuberous rhizomes or corms, a rosette of hairy leaves and axillary inflorescences. Sepals six, the outer three enclosing the inner in bud; stamens six. Ovary inferior, of three carpels.

A small family sometimes included in the Amaryllidaceae (by Heywood 1978) but probably warranting recognition.

Hypoxis angustifolia Lam. [46B]

A herb, 15 to 60 cm tall, growing from a truncate corm 1.5 cm in diameter. Leaves pubescent, much longer than the inflorescence, 12 to 55 cm long and 1 to 5 mm broad. Inflorescences 1 to 5, 2- to 4-flowered and up to 23 cm long. Flowers 1.6 to 1.8 cm in diameter, pubescent and greenish on the outer surface and yellow within.

It is widespread in our area from 1 300 m in the Ruaha National Park, where it grows in grassland with scattered bushes, up to 2 550 m on the Matamba Pass which leads to the Kitulo Plateau. Around Sao Hill it can be found in wet meadows growing among grasses at 2000 m while to the south-west it can be found at 1 800 m in the open woods of the Chimala escarpment.

This species is widespread in the savannas and montane grasslands of tropical Africa and is also found in Madagascar.

Hypoxis urceolata Nel [46C]

A larger coarser plant than the previous species, growing from a truncate bulbous base some 4 cm in diameter. Leaves long, pubescent, variable in length, up to 70 cm long and 8 to 12 mm broad. Inflorescence usually 4- to 7-flowered, 15 to 30 cm long, borne on a slender hairy stalk, the flowers, 2.6 to 3 cm in diameter, greenish and hairy on the outside, yellow within.

It is uncommon in our area but can be seen growing in the grasslands of the Mmemya Highlands in Ufipa at 2 100 m and in Njombe District between Lisitu and Lugalawa at 2 000 m flowering from September until December.

Elsewhere in Africa it is widely distributed from Uganda and Kenya south to Mozambique.

Plate 47. A, *Commelina forskalaei;* B, *C.subulata;* C, *Cyanotis longifolia*

COMMELINACEAE
THE SPIDERWORT FAMILY

Plate 47

Annual or perennial herbs commonly with creeping or ascending, jointed, succulent stems. Leaves entire, with a closed basal sheath. Flowers bisexual, usually regular, generally with a perianth of three free green sepals and three free coloured petals although, in some species, one of the petals is greatly reduced. Stamens in two series of three, one series often being sterile and modified to form staminodes. Ovary superior, commonly forming a thin-walled dehiscent capsule.

Some genera are ornamental and grown in cultivation, for example, *Zebrina, Tradescantia* and *Commelina.*

Commelina forskalaei Vahl [47A]

A mat-forming herb with procumbent sparsely hairy stems, up to 40 cm long, rooting at their nodes. Its leaves are yellow-green, up to 5.5 cm long and 1.2 cm broad. One or two bright blue flowers, about 1.5 cm in diameter, with prominent yellow anthers emerge from a boat-shaped bract about 1.1 cm long.

It is very widespread at lower altitudes from sea level near the coast around Dar es Salaam up to about 1 600 m in our area. It is particularly common in the Ruaha National Park and can be seen there and on the Great North Road north of Iringa growing in open *Commiphora* woodland and in flower from December until March.

Elsewhere it is widespread in Africa from Sudan and Ethiopia south to Zimbabwe and also in the Yemen where Forsskäl first collected it.

Commelina subulata Roth [47B]

A weakly branching annual herb 15 to 30 cm tall. The sparsely pubescent suberect leaves, usually 5 to 10 cm long, are striped with purple at the base. The bract which subtends the flowers is also purple-striped at the base and 8 to 9 mm long. The flowers, about 1 cm across, are apricot-orange or pale brown.

It is found in the open woodlands, amongst rocks and by streams up to 2 400 m in our area. It grows on the Usanga flats near Utencile at only 1 000 m while in Ufipa at Sumbawanga it grows in damp soil among rocks by streams at 2 400 m. It is not uncommon in damp sandy pockets in woodland, for example, at Ikuyu near Irambo in the Poroto Mts. It flowers from January until March in the area.

Elsewhere, this species is widespread from Ethiopia southwards to Zimbabwe.

Cyanotis longifolia Benth. [47C]

A very hairy plant with fleshy erect stems up to 40 cm long and fleshy swollen roots. The nodes, leaf- and stem-bases are commonly purple or crimson. The fleshy leaves reach 18 cm long and about 0.8 cm broad, but the upper leaves are much shorter. The fluffy azure-blue flowers, about 1 cm across, emerge from green, purple-tinged bracts. The central mass of blue filaments, covered with blue hairs and topped by yellow anthers, surrounds the central white style.

It is found at intermediate altitudes in the highlands up to about 2 200 m. On the Malonje Plateau it can be found growing in long grass in damp areas by streams while in the Poroto Mts it grows on the north slopes in grassland between bamboo thickets. It flowers from December until March.

Elsewhere in Africa, it is widespread from sea-level upwards from the Sudan southwards to South Africa.

ARACEAE
THE ARUM FAMILY

Plate 48

Mostly herbaceous plants with a great variety of vegetative habit but predominantly with aerial stems and underground rhizomes or tubers. The floating water plant *Pistia* is a member of the family. Many contain a milky or watery latex and crystals of calcium oxalate which are poisonous. Leaves simple or compound. Inflorescence characteristic, consisting of a large spathe enclosing or subtending a spadix on which numerous small flowers are borne. Flowers bisexual or unisexual, in the latter case both male and female flowers on the same spadix, the former above the latter. Four to six perianth segments are present only in bisexual flowers. The fruits are berries which are often brightly coloured.

Species of *Colocasia* (taro), *Xanthosoma, Alocasia, Amorphophallus* and *Cyrtosperma* are grown throughout the tropics for their edible starchy tubers. *Zantedeschia aethiopica* is widely cultivated for its attractive inflorescences which are often used as cut-flowers.

Amorphophallus abyssinicus (A.Rich.) N.E.Br. [48A]

Amorphophallus is a genus of aroid which contains some spectacular species with attractive, highly dissected leaves and inflorescences which can often be seen and certainly smelt from a considerable distance.

The solitary leaf of *A.abyssinicus* illustrated here reaches 60 cm tall and is highly dissected, each of the three main sections of the leaf being commonly twice pinnate. The leaf arises from a large flattened subterranean tuber about 10 cm in diameter. Both the leaf and inflorescence are flecked with maroon or dark green. However, the inflorescence appears well before the leaf develops. At the apex of the inflorescence stalk is a broad funnel-like spathe, 14 to 23 cm long, which is commonly rich velvety maroon within. This surrounds a maroon spadix 13 to 22 cm long.

A.abyssinicus is a widespread species in tropical Africa south to Zimbabwe. It is found at moderate altitudes up to about 1 650 m in our area in grassland which is seasonally burnt or under *Acacia* or *Brachystegia* trees in open woodland. It commonly flowers in November and December.

In our area it can be seen east of Sumbawanga on the road from Mpui to Kwela and in the Ruaha National Park.

Zantedeschia albomaculata (Hook.) Baill. [48B]

This attractive species is closely related to the well-known and common Arum or Madonna lily, *Z.aethiopica.* Like that species it is a denizen of swampy areas

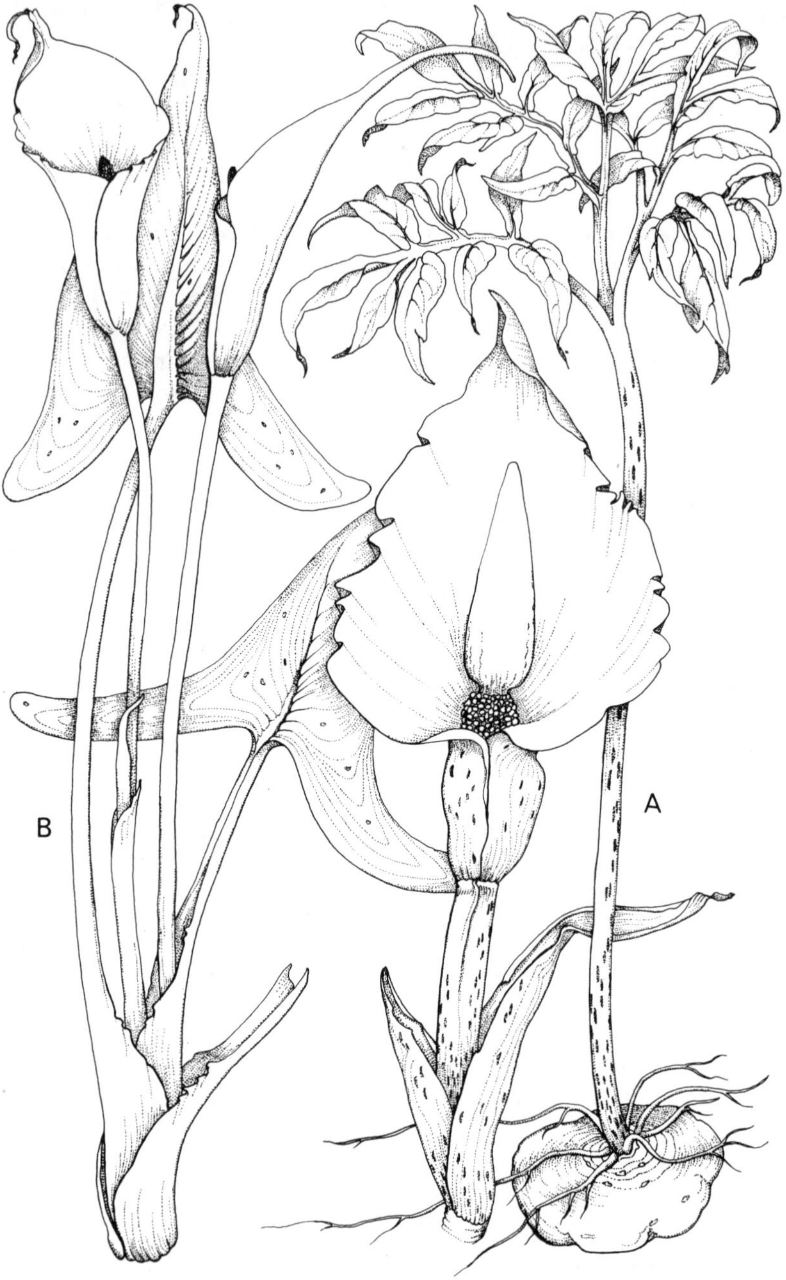

Plate 48. A, *Amorphophallus abyssinicus;* B, *Zantedeschia angustifolia*

where it grows in deep mud often amongst tall grasses and other marsh plants.

Z.albomaculata reaches about 60 cm tall but is often shorter, each plant bearing about three spear-shaped leaves with blades 11 to 40 cm long. The leaves are commonly dark green and are spotted with white. In flower, *Z.albomaculata* can be seen from a distance, its lemon-yellow funnel-shaped spathes standing out like flags. Each spathe is 8 to 12 cm long borne at the top of a 20 to 30 cm long stalk. Within the spathe is a 6 cm long orange-yellow spadix.

It is a local species but the inflorescences can be seen in December and January in marshy areas and swamps at Mbozi and near the Kuku River and on the Mwimbi to Sopa road near Sumbawanga at altitudes between 1 400 and 1 800 m.

Plate 49. A, *Dierama pendulum;* B, *Hesperantha petitiana;* C, *Radinosiphon leptostachya*

IRIDACEAE
THE IRIS FAMILY

Plates 49, 50 and 51

Perennial herbs growing from corms (e.g. *Gladiolus*), rhizomes or bulbs. Leaves narrow and linear, commonly arranged in a fan. Flowers bisexual, with six perianth segments arranged either regularly or irregularly and sometimes united into a tube below. Three stamens are arranged opposite the outer three segments. Ovary inferior and styles commonly branched or even petaloid (e.g. in *Moraea*); fruit a capsule.

The family contains many ornamental genera often found in cultivation notably *Gladiolus, Moraea, Freesia, Ixia, Crocosmia, Iris* and *Crocus.*

Dierama pendulum (L.f.) Bak. [49A]

A robust densely tufted perennial, up to 1.8 m tall, growing from a fibrous sheathed corm, 2 to 4 cm in diameter, with leaves 45 or more cm long and 0.6 to 1.5 cm broad. Flowers about 3 cm long, delicate, nodding, pink, purple or mauve, borne in a panicle on wiry peduncles and subtended by large purple-flecked bracts.

Throughout the grasslands of the Southern Highlands above about 2 000 m, it is a common and showy species flowering over a long period from about September to May. It is found from about 2 200 m to the summit of Mbeya Mt and similarly also on the other mountains of the region, for example, on Imagi Mt, NE of Iringa, and on Malonje Mt in Ufipa. On Kilimanjaro, it grows up to about 3 400 m and elsewhere it is widespread from Ethiopia south to Cape Province of South Africa.

Hesperantha petitiana (A.Rich.) Bak. [49B]

A delicate herb, 15 to 65 cm tall, growing from a scaly corm with a dull sepia-brown speckled stem and subulate leaves up to 32 cm long and 2 to 8 mm broad. Flowers 1 to 6, about 2.5 cm long, pink to pale mauve, red-flecked on the outer surface and with a yellow spot in the throat and mauve-brown anthers.

It is locally common from the southern Sudan and Ethiopia southwards to eastern Zimbabwe growing in montane grassland from about 2 000 to 3 400 m (e.g. on Mt Kilimanjaro). In our area it has been collected only infrequently, for example, on Imagi Mt, 80 km NE of Iringa, growing in montane grassland at 2 400 m and on the Kitulo Plateau at 2 000 m in similar habitats flowering in January and February.

Plate 50. A, *Gladiolus atropurpureus;* B, *G.dalenii;* C, *G.laxiflorus;* D, *G.unguiculatus*

Radinosiphon leptostachya (Bak.) N.E.Br. [49C]
A perennial herb, 20 to 45 cm tall, growing from an ovoid tunicate corm, about 1 to 1.6 cm long and 0.8 to 1.2 cm in diameter, and with slender grass-like leaves up to 30 cm long and 5 to 8 mm broad. Flowers 3 to 11, 1.5 to 4.5 cm long, bright mauve or purple.

It is widespread in our area, above about 1 250 m, growing on rocky ground in open woodland and grassland. It can be seen in flower in February in the turf at the summit of Mmemya Mt in Ufipa and around Sao Hill in January growing in peaty soil amongst lichen-covered granite boulders at 2 050 m.

This species is widely distributed elsewhere in southern Africa as far south as the Transvaal and Lesotho.

Gladiolus atropurpureus Bak. [50A]
An erect perennial herb, up to about 75 cm tall, growing from a fibrous-sheathed corm about 2 cm in diameter with sheathing leaves up to 20 cm long and 6 mm broad. Flowers 4 to 10, secund, 2.5 to 3 cm long, pale to dark purple or maroon with a paler base.

This attractive species which flowers in November and December is common in the grasslands around Mbozi at 1 650 m and Mufindi at 1 800 to 1 900 m. In Ufipa, it grows in grassy places in open woodland on the Namwele escarpment at about the same altitude. It reaches the northern limit of its range in Tanzania which extends southwards to Zimbabwe.

Gladiolus dalenii van Geel (syn. *G.psittacinus* Hook. & *G.natalensis* (Eckl.) Reim ex Hook.f.) [50B]
A robust plant, 50 to 150 cm tall, growing from a corm, 3.5 cm in diameter, with leaves 40 to 120 cm long and 0.7 to 3 cm broad. Flowers 3 to 9, very variable in size and colour, 3.5 to 6 cm long, white, yellowish green, orange, pink, red or brown, often flecked with reddish or brown.

It is by far the commonest *Gladiolus* in the area growing in grassland and flowering throughout the rainy season. For example, it can be seen in Ufipa at 2 100 m on Chala Mt in December, at 2 000 m around Sao Hill in March and at 2 400 m on the Kitulo Plateau in May. Most tribes have a name for it; the Wafipa call it 'kinkwesikwesi', the Wamatila 'ishilungu' and the Wanyiha 'ishinwinzimwinzi'.

It is widespread throughout tropical and southern Africa.

Gladiolus laxiflorus Bak. (syn. *G.tritonioides* Bak.) [50C]
A perennial herb, up to 120 cm tall, from a depressed brown-scaled corm about 3.5 cm in diameter. Leaves, strap-shaped, more or less falcate, often red-veined, up to 35 cm long and 0.8 to 1.3 cm broad, appearing with or shortly later than the inflorescence. Flowers 4 to 16, 2 to 4.5 cm long, pink or lilac, darker veined and with a yellow mark in the middle of the dark reddish or mauve throat.

It is widespread in the Southern Highlands from about 1 700 to 2 650 m flowering from October until February. For example, it grows in a swamp in Ukinga about 6 km before Kisinga on the road to Kitulo together with the orchids *Satyrium robustum* and *Centrostigma occultans.* Other localities where it can readily be seen are near Sao Hill, in the Lihogosa swamp near Njombe and at Mwela in Usafwa.

It is also found in Zambia and Angola.

Gladiolus unguiculatus Bak. (syn. *G.flexuosus* Bak.) [50D]
Plants, 25 to 70 cm tall, growing from a fibrous-sheathed corm, 1.5 to 3 cm in diameter, with leaves, up to 16 cm long and 0.4 cm broad, absent at flowering time. Inflorescences secund, 4- to 14-flowered; the flowers white to mauve, often with deeper marks on the lower petals, 2.5 to 3 cm long.

A species found in grassland or open woodland flowering from December until February in our area where it is widespread. For example, it grows in Mufindi in open grassy places in the forest by the Kigogo River at 1 900 m, on Mbeya and Rungwe Mts from 1 600 to 2 300 m, on the Kitulo Plateau at 2 400 m and in Ufipa down to 1 500 m.

Elsewhere it is widespread throughout tropical and southern Africa.

Moraea bella Harms. [51A]
Species of the genus *Moraea* resemble those of the north temperate genus *Iris* so often grown in gardens throughout the world. This slender attractive species commonly varies from 40 to 65 cm tall and grows from a 1.5 cm diameter corm which is covered by pale reticulate fibres. The slender channelled leaf is 2.5 to 6 mm broad but exceeds the inflorescence in length. From the dry brown spathes at the apex of the stem emerge one or two pale yellow flowers, commonly darker veined and speckled, with the outer segments 4.5 to 5.5 cm long.

It is widespread in the Southern Highlands from Dabaga westwards to Ufipa and flowers from January until July but with a peak of flowering in April. It can be seen flowering in March in grassland at the top of the Matamba escarpment by the road. In most places it is found in wet places such as in marshes and swamps. It is widespread outside Tanzania in Malawi, Zaire, Zambia and Mozambique.

M.verdickii, also found in our area in similar situations around Dabaga and north of Mbeya on the road from Makongolosi to Kampikatoto, is closely related but generally flowers earlier and has a larger flower in which the outer floral segments are each up to 10 cm long.

Moraea callista Goldblatt [III A]
Although one of the most attractive species of the genus, *M.callista* has only recently been described due, no doubt, to its very restricted distribution for it has so far only been seen once in the Uluguru Mts and a few times on the Kitulo Plateau.

Plants range from 30 to 70 cm tall and have two leaves both being longer than the inflorescence. The flowers emerge from herbaceous spathes and are attractively coloured white with a blue or mauve centre. The flowers are 6 to 7 cm across.

We have seen it in a few places on the Kitulo Plateau growing in rocky areas of the plateau margins and on peaks such as Mtorwi and Ishinga where it grows scattered amongst rocks (Buanji sandstone) a hundred metres or so below the summits. It also grows by the roadside in rocky ground on the pass leading from Kitulo to Matamba at about 2 700 m. It usually flowers in January and February.

Moraea schimperi (Hochst.) Pic.-Serm. [51B]

This widespread species ranges from Zimbabwe northwards to Ethiopia and Sudan and westwards across to Nigeria.

It is a most distinctive species, 20 to 40 cm tall, with the flowering stem emerging with or before the new season's leaf appears from a 2 cm corm. The prophylls, bracts and spathes are distinctively large and rich brown in colour and the flowers have 4 to 6 cm long blue-purple segments with yellow-white nectar guides. The leaf when fully developed is much larger than the inflorescence and varies from 9 to 15 mm broad.

They produce a wonderful display in the Southern Highlands flowering in their thousands after the grasslands have been burnt. This happens on the Kitulo Plateau in September when for kilometre after kilometre the undulating country is dotted with this species growing amongst many other treasures such as *Disa hamatopetala, Scilla rigidifolia* and *Cyrtanthus* and *Gnidia* species.

Its flowering time varies from August until well into the rain in December. The Wasafwa call it 'malungulungu'.

Moraea tanzanica Goldblatt [51C]

The two common yellow-flowered *Moraea* species of the Southern Highlands are *M. tanzanica* and *M. verdickii.* In the former, the leaf seldom reaches beyond the middle of the spathes whilst, in the latter, it exceeds the inflorescence.

M. tanzanica is a most attractive, albeit small, species. It seldom reaches 35 cm tall arising from a small 1.5 cm diameter corm. The leaf ranges from 15 to 25 cm long. The solitary flower is bright canary yellow and the 4.5 to 5.5 cm long outer segments characteristically droop.

It occurs on Rungwe Mt, Mbeya Peak and on the Kitulo Plateau. On the latter we saw it on an unforgettable day in January when we climbed Ishinga Mt from the village of Mwela which at 2 800 m must be one of the highest inhabited villages in Tanzania. The grasslands covering the lava area are not rich in flowers and we did not see *M. tanzanica* until we approached the plateau edge where a thin rocky soil replaced the lava. Here we also saw the rare *M. callista* and, among the rocks, undescribed dwarf species of *Aloe* and *Ceropegia.*

Plate 51. A, *Moraea bella;* B, *M.schimperi;* C, *M.tanzanica;* D, *M.thomsonii;* E, *Romulea campanuloides*

Moraea thomsonii Bak. [51D]

Ten of the 24 tropical African species of *Moraea* grow in the Southern Highlands of Tanzania and of these *M. thomsonii* is probably the most widespread and commonest, ranging from Ethiopia to South Africa.

It is a delicate plant, 12 to 30 cm tall, growing from a 1 to 2 cm diameter corm covered in dark brown reticulately fibrous tunics. The leaf is either lacking or quite dry by flowering time and the flowers are smaller than in the other species described here, the outer petals being only 1.5 to 2 cm long, pale lilac-blue in colour with yellow-orange nectar guides.

A typical locality for this delightful species is on the rocky hill of Majembe (2 000 m), overlooking the Songwe valley in the western Bundali Hills. Early September at the height of the dry season with only the dry dessicating east wind blowing over the hillsides is an unlikely time to look for flowers. However, numerous plants of *M. thomsonii* can be seen here in full flower along with the pink flowers of the orchid *Polystachya dendrobiiflora,* the bright red of *Hibiscus rhodanthus,* the red and yellow of *Gnidia kraussiana* and the yellow of *Hirpicium angustifolium* peeping out from the ground in dense clumps. Elsewhere in the area *M. thomsonii* can be seen in flower as late as November mostly above 2 000 m.

Romulea campanuloides Harms. [51E]

Although not uncommon in the higher grassland areas of the Southern Highlands, this pretty iris relative is often overlooked because of its small size.

It grows from a small bulb and produces a solitary flower on a slender stalk up to about 15 cm long. The one or two leaves are grass-like and up to 30 cm long. The rosy-mauve or purple star-like flowers are eye-catching and about 1.5 to 2 cm across. In the centre are three yellow anthers surrounding the stigma which is divided into six branches.

Mary Richards found this delightful species in several places on Rungwe Mt, on the higher levels in Ufipa and on the Kitulo Plateau. A typical locality in the latter area is on the slopes of Mtorwi Mt where it flowers in the short turf in December and January. It is readily confused where it occurs together with *Hesperantha petitiana* (A.Rich.) Baker, also in the Iridaceae, which can be distinguished by its flowers borne in a raceme and by having only three stigma branches. It also flowers later in the year in May.

Plate 52. A, *Albuca angolensis;* B, *A.pachychlamys;* C, *Anthericum cooperi;* D, *A.subpetiolatum*

LILIACEAE
THE LILY FAMILY

Plates 52 to 55

Mostly herbs with swollen storage organs such as bulbs, corms, rhizomes or fleshy swollen roots but a few genera such as *Aloe* are evergreen succulents. Leaves very variable, commonly one, linear and parallel-veined but in *Gloriosa* the leaves end in a long tendril while in *Asparagus* they are often reduced to scales. Flowers regular, bisexual, with six more or less equal perianth segments in two whorls of three or united into a tube (as in *Kniphofia*). There are usually six stamens arranged opposite the perianth segments; ovary superior. Fruit either a dry capsule or less frequently a berry (as in *Asparagus*).

An economically important family both agriculturally, e.g. onions, asparagus, leeks and chives, and horticulturally, e.g. lilies, aloes, red-hot pokers (*Kniphofia*) and *Gloriosa.*

Albuca angolensis Welw. ex Bak. (syn. *A.chlorantha* Welw. ex Bak.) [52A]
Several species of the lily genus *Albuca* grow in our area, mostly at lower altitude. They are recognised by their white or yellow and green flowers which do not open widely. One of the commonest and most attractive of these, *A.angolensis,* stands 30 cm to 1 m tall growing from a depressed bulb 2.5 to 4 cm in diameter. It has several long leaves up to 60 cm long and 2 cm broad. The several to many-flowered inflorescence is produced from March until May. The flowers are yellow, the inner segments, 2 to 2.6 cm long, having central green keels. Inside the flowers, the filaments are green and the central style is whitish green. At the base of each flower stalk is a bract, 2 to 3.3 cm long.

A.angolensis can be seen throughout the region. It grows in woodland and in grassland with scattered shrubs by the main Iringa to Mbeya road near Sao Hill and by Lake Ngwazi at Mufindi. It is also common in the Poroto Mts on the south slopes around Isongole between 2 000 and 2 250 m, on the Kitulo Plateau and in Ufipa in the Mpui area. It flowers from April until May.

Elsewhere in Africa it is widespread in a belt across south-central Africa as far south as Zimbabwe.

Albuca pachychlamys Bak. [52B]
This little lily, a common and widespread species in South and south-central Africa, reaches the northern extent of its distribution in southern Tanzania and adjacent Malawi.

It sends up a few- to many-flowered unbranched inflorescence from a white onion-like bulb 3.5 to 5.5 cm in diameter. The leaves are fleshy and up to

30 cm long and 1 cm broad. The attractive flowers, 1.5 cm long, have each white segment marked with a central grey-green streak. It grows in grassland at about 1 600 m in our area and is prominent only in flower when the first rains bring life to areas that have been burnt off. A typical locality is in burnt grassland in the Ukinga highlands near Tandala where it flowers in November in association with the tiny *Hypoxis dregei* and *Gnidia kraussiana.*

Anthericum cooperi Bak. (syn. *A.zanguebaricum* Bak.) [52C]
With the first showers of the rainy season falling on the often burnt landscape in October many flowers push their way out of the ground. Amongst these is the lily *A.cooperi,* a small plant 7 to 35 cm tall, growing from a bulb almost completely hidden by copious fibrous roots. It has six linear leaves often up to 20 or 30 cm long and 0.5 to 1.6 cm broad. The inflorescence is a short cylindric head of about ten flowers clustered together at the apex. The flowers are white with yellow anthers and are borne on short stalks subtended by blackish brown bracts. The petals, about 7 to 15 mm long, are thin in texture.

It is a common plant throughout the region. We have seen it in early October on the rather rocky hillsides around Ibaba in the western Undali Hills at about 2 250 m. This is a marvellous area for flowers throughout the year. Rocky hillsides fall away steeply on three sides to the woodlands below. *A.cooperi* grows here in association with two *Hypoxis* species, *H.angustifolia* and *H.multiflora,* and the rare miniature white-flowered *Trachyandra asperata.* Elsewhere in the area it can be seen in Ufipa at Mbisi flowering in November in grassland and at Malonje at 2 600 m on rocky hillsides. It is also common on the Mbeya Range and can be seen by the roadside on the Chunya road at 2 100 m in January.

A.cooperi is widespread in Africa from Kenya south to South Africa.

Anthericum subpetiolatum Bak. [52D]
This is one of the many white-flowered bulbous species of lily found in our area. It is a small plant growing from a horizontal rhizome swollen at intervals to about 1 cm in diameter. Its grey-green leaves reach 25 cm long and 2 cm across. The delicate inflorescence is generally shorter than the leaves and carries several star-shaped flowers with white petals 1 to 1.5 cm long and yellow anthers.

Two varieties of this pretty species grow in our area. The typical variety with several narrow leaves is common in Ufipa, in the Ruaha National Park and around Iringa from about 750 m upwards flowering in November and December.

Var. *monophyllum* (Bak.) Hanid, distinguished by its single broader leaf, is commonly found at higher altitudes up to over 2 100 m growing in woodland and swamps. We have found it on the Chimala escarpment and in the Umalila mountains. In the latter area much of the original grassland has been cultivated during the last two decades as the soils are very fertile and the population is increasing rapidly. However, the roadsides are still rich in flowers and, in November here at about 2 100 m, var. *monophyllum* can be seen in flower in the grassy banks between Santilya and Isuto.

A.subpetiolatum is widespread in East and South-central Africa from Ethiopia southwards to Zimbabwe.

Aloe chaubardii Schonl. [53A]

Several species of *Aloe,* a succulent-leaved genus of the lily family, are found in the Southern Highlands area. The four described and illustrated here are those most frequently met with in the Mbeya region.

A.chaubardii is a stemless or short-stemmed plant which commonly suckers so that clumps of several shoots are frequently found. Each shoot has about 20 brittle glaucous or greyish green leaves which are obscurely lined but rarely spotted with white above. The leaves reach 50 cm long and 10 cm broad at the base and have marginal flesh- or straw-coloured prickles. The branched inflorescence, somewhat pyramidal in outline, reaches 60 to 80 cm tall. The curved flowers, subtended by bracts 3 to 5 mm long, are coral-pink or pale vermillion with orange-yellow tips and 3.5 to 4 cm long.

A.chaubardii is widespread in South-central Africa from Zimbabwe north to southern and western Tanzania and southern Zaire. In our area it grows in similar situations to *A.christianii,* on flat rocks on slopes in woodland at about 1 500 m. We have seen it near the Iringa to Mbeya road 60 km E of Mbeya by the Kimani River where it flowers from July until September.

Aloe christianii Reynolds [53B]

This large *Aloe,* a glorious sight when in full flower, grows either singly or in small clumps and has a short or rarely longish stem to 1 m long which is then generally decumbent. Each stem carries a dense rosette of 30 to 40 leaves, 30 to 50 cm long and up to 10 cm broad at the base. They are dull bluish green in colour, obscurely lined on the upper surface and have smooth pale brown marginal teeth 1 to 2 cm apart. From the apex of the stem arises a large inflorescence, 2 to 2.75 m tall, with up to 30 branches, each bearing 40 to 50 flowers in a 25 to 30 cm long head. The flowers are coral-red, rounded at the base and from 3.5 to 4.5 cm long.

This part of Tanzania is almost the northernmost limit of its range which extends south to Zimbabwe. Mary Richards collected it in the Usangu forest near Mole in SE Ufipa in August 1960. She found it growing on a great outcrop of rock at 1 800 m in association with another species, *A.mzimbana* and other plants such as *Crotalaria natalitia, Senecio syringifolius, Ipomoea involucrata, Euphorbia depauperata* and *Gnidia kraussiana.* Elsewhere, it is quite common just east of Mbeya by the Iringa road at between 1 500 and 2 150 m growing in grassland and flowering from February until June.

Plate 53. A, *Aloe chaubardii;* B, *A.christianii;* C, *A.nuttii;* D, *A.lateritia*

Aloe lateritia Engl. [53D]

Another *Aloe* with a branching inflorescence is *A.laterita,* the most widely distributed species in the Southern Highlands growing at all altitudes to the very highest mountain tops.

It has a basal rosette of 16 to 20 leaves on a very short stem. The leaves are dull to bright green, copiously marked with elongated white spots and streaks and have strongly toothed margins, the teeth being light brown and 10 to 15 mm apart. From the centre of the stem emerges a stout branching inflorescence, up to 1.25 m tall, each branch of which has a capitate or cylindrical many-flowered raceme 8 to 25 cm long. The 3.5 to 3.8 cm long flowers are bulbous at the base and constricted immediately above. They are glossy and bright scarlet, orange or coral red in colour with a paler mouth.

A.lateritia begins to flower in April at lower altitudes on the plains around Mbeya and rather later in the mountain areas where flowering can go on until August or even later. It is the most obvious flower on the lowland plains of the Usangu during April and June. Other places where it can be easily seen are around Iringa and Sao Hill where it grows in bushy grassland at 2 000 m and above Mbeya about 25 km along the Chunya road in grassland at 2 350 m. The juice is used everywhere as a medicine to help childbirth and for magic and most tribes have a name for it. The Wakinga call it 'lyusi', the Wanyakyusa it 'ibugubugu' and the Wahehe 'litembwetembwe'. The name 'itembwe' is in fact widespread amongst the tribes for all aloes.

A.lateritia is widespread in East Africa and is also found in Zaire.

Aloe nuttii Bak. [53C]

When seen from a distance this plant with its simple unbranched inflorescence looks rather like a red-hot poker. It has one to three short-stemmed shoots, each with 16 to 20 succulent but somewhat grass-like leaves, 40 to 50 cm long and 1 to 2 cm broad, above the rather broader base. The leaves are green more or less obscurely lined above and with a few small white spots at the base. Each shoot has one or two inflorescences, 60 to 80 cm long, bearing many flowers in a 10 to 20 cm long and 9 to 10 cm wide head at the apex. The flowers, 4.2 cm long, orange-red or salmon-pink, are subtended by large flesh-coloured bracts 1.5 to 2 cm long.

A.nuttii is widespread in the Southern Highlands, at higher altitudes mostly above 1 000 m, where it flowers from January until June. Mary Richards collected it several times; for example, in January 1961 she found it on the Kitulo Plateau at 2 400 m and, in April the same year, on Kito Mt in the Ufipa Highlands. On the Kitulo Plateau it grows in wet areas with *Gunnera perpensa, Rubus porotoensis, Clerodendrum johnstonii, Indigofera mimosoides* and *Bartsia abyssinica.* The Wasafwa call it 'wtembwe'.

Elsewhere in the Southern Highlands, it can be seen near Iringa and Njombe as well as on the Mbeya and Poroto Mts. In Tanzania it is at the northern limit

Plate 54. A, *Kniphofia grantii;* B, *K.kirkii;* C, *K.thomsonii*

of its range which extends through Malawi, Zambia and southern Zaire to Angola.

Kniphofia grantii Bak. [54A]
A variable species in height ranging from 30 cm to over 1 m tall when in flower. It often forms clumps of up to three-shoots covered at the base by many coarse erect fibres, the remains of old leaves. The leaves have finely saw-toothed margins and vary from 6 to 75 cm long and 1.2 to 3.8 cm broad. The conspicuous inflorescence bears a 5 to 15 cm long, 4 to 6 cm in diameter, ovoid to cylindric dense cluster of many flowers. The flowers are variable in size being 1.5 to 3.9 cm long and are commonly yellow-green or greenish but rarely are orange-yellow. The bracts are hidden amongst the flowers and range from 8 to 23 mm long.

K.grantii (which is also known under a number of synonyms such as *K.zombensis* Bak., *K.flavovirens* Engl. and *K.mpalensis* Engl.) is relatively widespread in East and south-central Africa from Uganda south to Zambia and Malawi. It is not uncommon in our area over a wide range of altitude from about 780 to 2 700 m growing in grassland and in seasonally flooded marshes. It commonly flowers from September until May with the peak flowering occurring from October until January. *K.grantii* should be looked for in the wet marshes near Sao Hill, in the grasslands of the Kitulo Plateau and Poroto Mts where it is very common and again in hilly areas of the Mbozi and Sumbawanga districts.

Kniphofia kirkii Bak. (syn. *K.paludosa* Eng.) [53B]
One of the rarest red-hot pokers of the Southern Highlands and endemic to the Kitulo Plateau and Kipengere Range.

It is also one of the largest species forming clumps up to 1.3 m tall. The coarse leaves reach 1.8 m long, are 1.8 to 4 cm broad and have minutely saw-toothed margins. It produces densely many-flowered inflorescences 10 to 12 cm long and 7 to 9 cm in diameter. The spreading reddish-orange or coral-coloured flowers have yellow tips and become pendent with age. The flowers are commonly between 3.5 and 4.2 cm long and the bracts are 1 to 1.3 cm long.

It always grows with its feet in water and can be seen flowering between November and January in the streams of the Kitulo Plateau near the cattle station growing at altitudes between 2 500 and 2 800 m.

Kniphofia thomsonii Bak. [53C]
A widespread species in East Africa from Uganda southwards to the Southern Highlands of Tanzania.

It often forms a large plant with shoots, up to 2 m tall, growing in small clumps from a rhizome which is bright yellow when cut. The leaves are commonly shorter than the inflorescence at flowering time but later on may reach

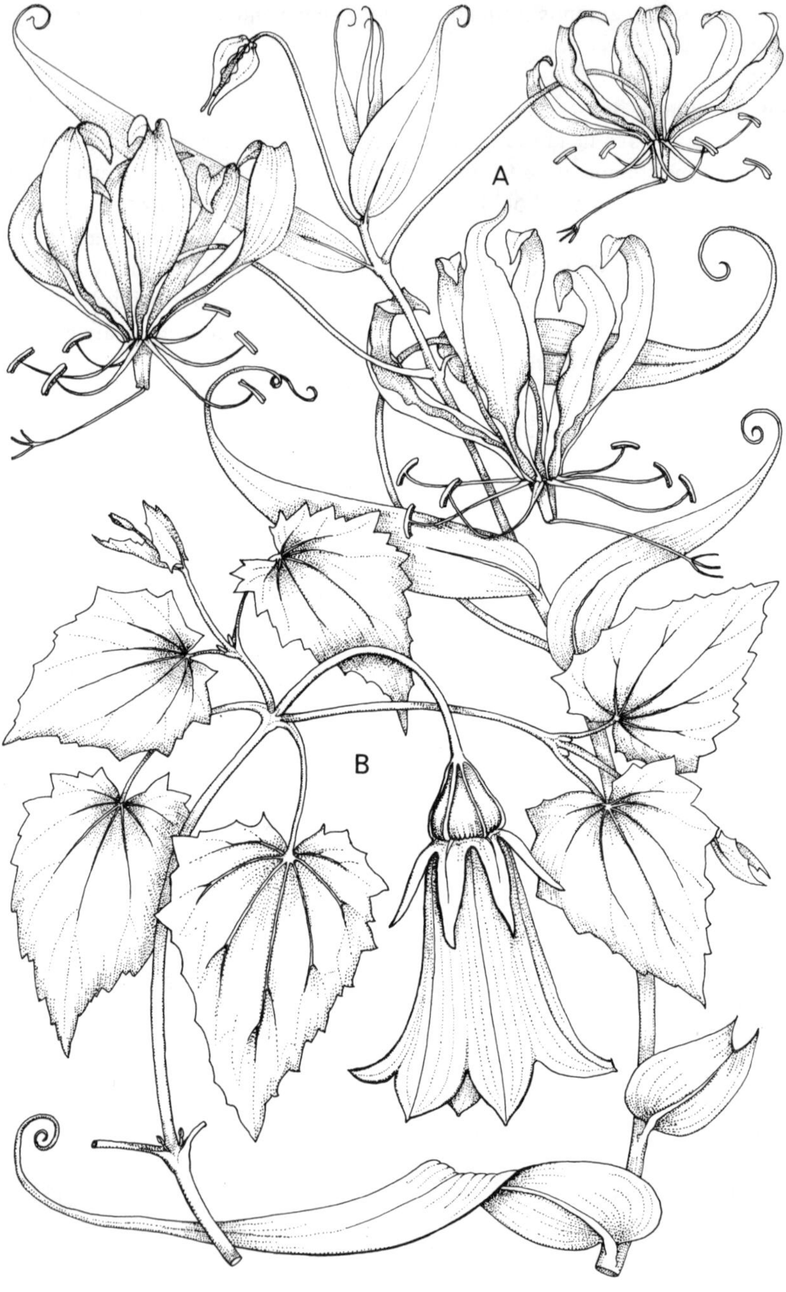

Plate 55. A, *Gloriosa simplex;* B, *Canarina eminii*

90 cm long and 0.7 to 2.5 cm broad. In contrast with *K.grantii* and *K.kirkii* the inflorescence is much longer, up to 40 cm long and 4 cm in diameter, and is relatively laxly many-flowered. The yellow, orange or red flowers range from 2.2 to 3.5 cm long and are subtended by 0.4 to 1.4 cm long bracts.

Two varieties of this species have been recognised: var. *snowdenii* with broader leaves and a longer inflorescence occurs only in Kenya and Uganda; var. *thomsonii* widespread in Zaire, Kenya and Tanzania is the only variety found in the Southern Highlands. It can be seen flowering in May near Mwakaleli on the banks of the Mwatesi River and also in the Mufindi area growing on the Lugoda Tea Estate. It is a widespread species found between 1 500 and 2 400 m altitude in the area. The Malila people call it 'susumba'.

Gloriosa simplex L. [55A]

This spectacular lily is probably the best known and loved of all the flowers of the Southern Highlands. It begins to flower around Christmas time and deserves its common name of glory lily for its flowers are truly a glorious sight.

G.simplex is a very variable species both in habit and flower size and colour. It grows in a wide variety of habitats from grassland, where it is a free standing plant often only 30 cm tall, to montane forest where it climbs into the trees and may have several metres long scrambling stems. It commonly climbs by means of tendrils at the apex of the lanceolate leaves whose blade reaches 14 cm long. The flowers are commonly yellow and red but both pure red and pure yellow forms are not uncommon.

Undoubtedly the best flowering forms are to be found in the wet mountain forests at about 2 000 m. It is a common sight in February in full flower by the roadsides around Mbeya growing in grassland or scrub and also in the Poroto Mts growing in forest at the same time of the year. The Wamalila people use an extract of the tuber to cure earache.

Canarina eminii Schweinf. [Campanulaceae] [IIA & 55B]

Just two species of this attractive genus of bell-flower occur in East Africa. *Canarina abyssinica* which is found from the Sudan and Ethiopia southwards to northern Tanzania may be distinguished by its long coiled petioles and coiled pedicels. *C.eminii,* the only species to be found in the Southern Highlands, has shorter uncoiled petioles and uncoiled pedicels. This latter species may be either epiphytic or less frequently terrestrial with erect to pendent stems up to several metres long. The leaves reach 10 cm long and are borne on a petiole less than 5 cm long. The large yellow to orange flowers are longitudinally veined with red and reach 7.6 cm long.

A typical locality is in the Irambo rain forest in the Poroto Mts. This lovely forest used to stretch from Irambo at 2 300 m down to 2 000 m and must have been a superb sight. However, the upper part of the forest has been felled for cultivation and there are pressures on the rest for both cultivation and timber extraction. *Aningeria adolphi-friedericii,* an immense tree up to 50 m tall towering above the canopy of other forest trees such as *Ilex mitis, Tabernaemontana, Prunus* and *Ekebergia* species, has great buttresses at its base so that 16 steps scarcely takes one around the base. From a few metres up this tree hang the long tresses of *C.eminii* bearing their numerous attractive orange bell-flowers. Elsewhere in the Mbeya area highlands, *C.eminii* is frequently encountered in the montane forests above 1 800 m. The local Safwa people call it 'lenyeshe' while the Malila tribe call it 'fundofundo'.

AMARYLLIDACEAE
THE DAFFODIL FAMILY

Plate 56

Herbs with a bulbous or less commonly a rhizomatous base. Leaves more or less linear and basal. Flowers bisexual, regular, made up of two whorls of three segments either free from each other or joined to form a tube (as in *Scadoxus*). In many genera a corona, commonly produced by the perianth segments, is present. Six stamens are present in two whorls of three. Ovary inferior and the fruit either a dehiscing capsule or a fleshy berry.

An horticulturally important family which includes the attractive genera *Crinum, Scadoxus, Zephyranthes* and *Nerine.*

Scadoxus multiflorus (Martyn) Raf. [IID & 56A]
Commonly known by its synonym *Haemanthus multiflorus* and more popularly as the fire-ball lily.

Its spectacular inflorescences are a familiar sight in Tanzania at the start of the rains. These appear before the leaves develop. On top of a 30 cm long stalk, which is spotted with red below, is borne a globose head, 15 to 20 cm in diameter, subtended by about 5 bracts and made up of many scarlet or vermilion flowers. The flowers are about 1 cm long and 1.8 cm across. The inflorescence emerges from a large bulb. Several large elliptic leaves appear later on at the side of the inflorescence. The basal sheaths of the sterile shoot are again red-spotted.

The fire-ball lilies are known as 'nyalendu' by the Usafwa people and 'nawaniga' by the Hehe. They are widespread and locally common from sea level up to 1 650 m. In our area they may be seen, for example, near Lake Sundu on the Kale Plain growing in overgrazed pasture amongst clumps of *Asparagus.*

Scadoxus puniceus (L.) Friis & Nordal (syn. *Haemanthus puniceus* L. & *H. goetzei* Harms). [56B]
This is a more typically montane species than the closely related *S. multiflorus.* Like that species, it also flowers before the leaves appear at the start of the rains. The inflorescences reach 40 cm tall with a stout 1 to 2 cm diameter stalk subtended by one or two red-spotted sheaths. At the top of the stalk is a large inflorescence, 8 to 10 cm across, almost enclosed by about eight large red bracts. These hide numerous flowers 6 to 7 cm long including their stalks. Their 3 cm long petals are red inside and red and white on the outside and have an ovary which is green. The inflorescences and large leaves, which appear later, emerge from a large bulb about 7 cm in diameter.

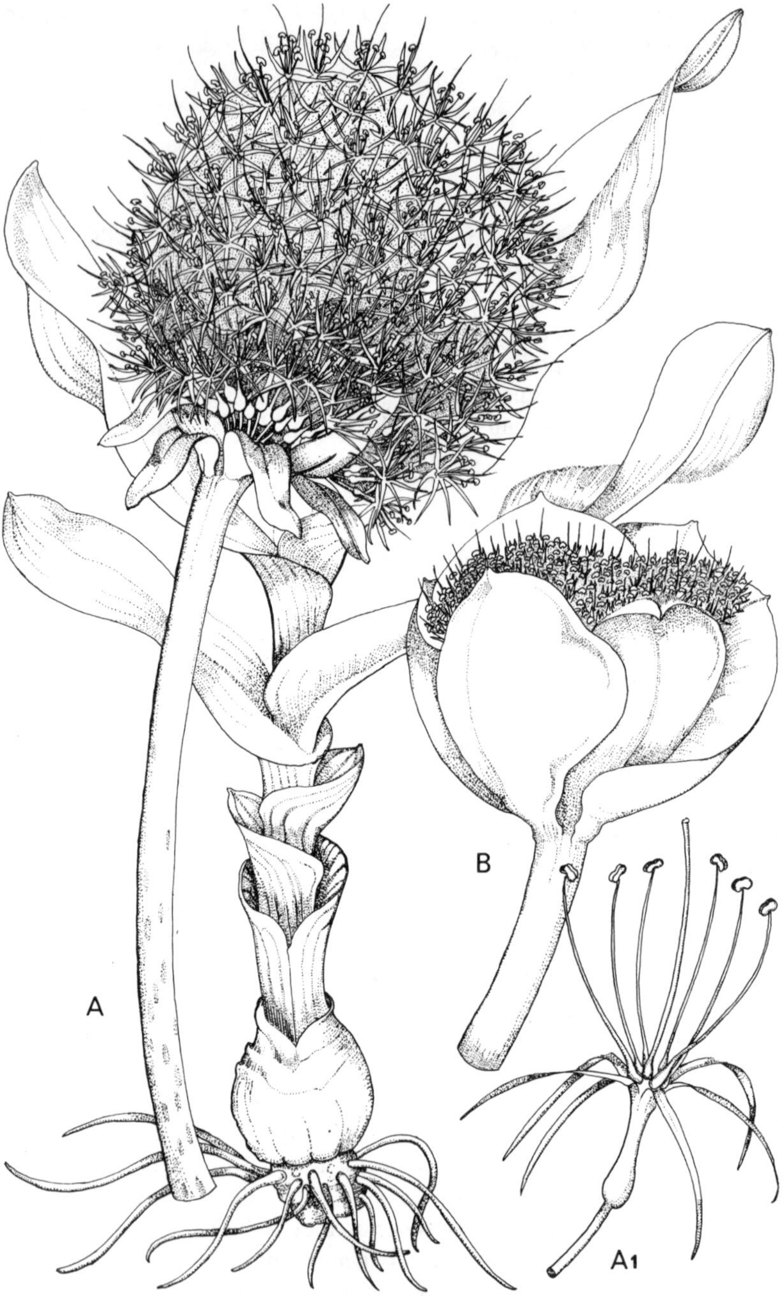

Plate 56. A, *Scadoxus multiflorus* (A1, flower); B, *S.puniceus*

S.puniceus is found in forests and woodlands between 1 500 and 2 300 m. It is not uncommon in the glens and valleys on the north slopes of Rungwe Mt, at the top of the Chimala escarpment and in the mountains around Sumbawanga, such as Kito Mt.

Plate 57. A, *Brachycorythis buchananii;* B, *B.pubescens;* C, *Brownleea parviflora;* D, *Corycium nigrescens*

ORCHIDACEAE
THE ORCHID FAMILY

Plates 57 to 72

A very large family of terrestrial or epiphytic (growing on tree branches and trunks) herbs. The terrestrial species have tubers or are rhizomatous and, although commonly green and leafy, several are saprophytic (e.g. *Epipogium, Eulophia galeoloides, Auxopus* and *Didymoplexis* in Tanzania). The epiphytic species commonly have stout roots and often leathery or fleshy leaves borne on long or short stems. In some genera, such as *Bulbophyllum,* the stems are swollen and are then called pseudobulbs. Flowers irregular, with three sepals and three petals, the lowermost petal commonly elaborated into a structure called a lip which is often spurred at the base and lies opposite the column on which is borne a solitary terminal anther (in all African orchids) and ventral stigmas. The pollen is borne in a discrete number of pollen masses.

The orchids are artificially divided in the following account into the terrestrial and epiphytic species. Orchids are popularly cultivated for their attractive and often fragrant blooms.

Brachycorythis buchananii (Schltr.) Rolfe [57A]

Some six species of the genus *Brachycorythis* occur in the Southern Highlands. All are attractive terrestrial species which enhance the beauty of the mountain grasslands and woods during the rainy season.

B.buchananii is one of the prettiest of these despite its tiny flowers. It forms neat plants 20 to 55 cm tall. Its leaves are small, up to 4.5 cm long and 1 cm broad, and the densely flowered inflorescence may be up to 14 cm long with many pink to purple or mauve flowers from 4 to 6 mm long.

This species is found throughout the Southern Highlands at moderate altitudes up to 1 950 m growing in swamps and in grassland. A typical locality is Ibililo near Kiwira in NW Unyakyusa. A rough road leading from Kiwira to the village crosses the Kiwira River and then a few kilometres further on can be seen the superb waterfall of the Marasusa where the Kiwira River hurls itself over a ledge of basalt framed between slopes clothed in the wild banana, *Ensete ventricosum.* Nearby are stands of the graceful palm *Raphia ruffia* and, in January and February, *B.buchananii* can be found growing in some abundance in the surrounding grasslands.

This species is also found throughout East Africa and elsewhere from Nigeria in the west south to Zimbabwe.

Brachycorythis pubescens Harv. [57B]

Few grassland orchids are hairy but two species of *Brachycorythis* are an exception both having stems, leaves and bracts which are densely softly hairy. The first of these *B.pubescens* is illustrated opposite. This is a large terrestrial orchid commonly reaching 85 cm tall. It has velvety leaves up to 6 cm long and 2.5 cm broad. The inflorescence is densely many-flowered and up to 35 cm long. Its flowers are pink or purple often with an orange centre and the lip varies from 5 to 10 mm long.

The only other species with pubescent leaves and stems is *B.velutina* Schltr. also found in our area, for example in Ulambya. It differs in having creamy or yellow flowers marked with brown or purple and a lip which is not bent but projects forwards or upwards and has a mid-lobe longer than the side lobes.

B.pubescens is widespread but never common in the Southern Highlands and throughout tropical and South Africa. It grows in upland grasslands and in open woodlands between 900 and 2 400 m. *B.velutina* is also a grassland species but is found only between 1 000 and 1 800 m.

Brownleea parviflora Lindl. [57C]

Brownleea is a small genus closely related to *Disa* but mostly with smaller less showy flowers. Most species are confined to South Africa except for *B.parviflora* figured here which is found throughout southern and tropical Africa as far north as Kenya and Cameroun.

It is readily identified by looking for the coarsely hairy sheaths at the base of the stems which reach 65 cm tall. The large leaves are 7 to 22 cm long and 0.5 to 2.5 cm broad. The whitish, lilac or violet flowers on close examination resemble a small *Disa* differing mainly in that their petals are united to the dorsal sepal. The spurred dorsal sepal reaches 5.5 mm long, the lateral sepals being somewhat shorter.

B.parviflora is a species of the higher altitude grasslands and pastures from 1 600 to 2 750 m but mostly above 2 000 m. A typical locality is the meadows leading down to Lake Dhambwe in the Poroto Mts. This lake is reached by a two-hour hike from Kikondo on the main Mbeya to Kitulo road. In late February the meadows there are alive with orchids including *Neobolusia stolzii, Satyrium crassicaule* and *S.comptum* and numerous other attractive flowers.

Corycium nigrescens Sond. [57D]

This is a South African species which has only recently been recorded from East Africa when it was discovered growing on the Kitulo Plateau in early March 1970. It is one of several species of South African affinity which just enter Tanzania at the northernmost extension of their range.

PLATE III

A | B
C | D | E

A *Moraea callista*, Ishinga, Kitulo Plateau, p. 20
B *Disa hamatopetala*, p. 203
C *Habenaria occlusa*, Kitulo Plateau, p. 213
D *Centrostigma occultans*, Mbozi, p. 201
E *Disa fragrans*, Kitulo Plateau, p. 201

It is a grassland species with its yellowish or brownish flowers almost concealed by the long bracts. The whole plant seldom reaches 40 cm tall and when picked it characteristically dries a black colour.

It is a rarity on the Kitulo Plateau but may be seen on the slopes of Chaluhangi Mt and in scattered localities towards the northern edge of the plateau. Chaluhangi is easily reached from the Kitulo farm. The view over Unyakyusa from the top is spectacular but the descent to the Lufirio River 1 000 m below is extremely difficult due to the steep slippery slopes covered with bamboo.

Calanthe sylvatica (Thou.) Lindl. (syn. *C.stolzii* Schltr.) [IVB & 58A]
One of the most handsome plants of the montane forests of the Southern Highlands, the large terrestrial orchid, *C.sylvatica,* grows in deep damp leaf litter often by streams. The 2 to 5 cm long pseudobulbs are covered by the bases of several large dark green pleated leaves up to 45 cm long and 13 cm broad. The pubescent inflorescence axis overtops the leaves bearing a few to many large flowers towards the apex. These are commonly rose-purple with a darker magenta lip although white-flowered forms are relatively common. The flowers are rather variable in size, up to 6 cm across, and have a spur, up to 3 cm long, at the base of the showy lip.

C.sylvatica is nowhere common in the Southern Highlands but where it occurs can grow in colonies of up to a few hundred plants. Such colonies can be seen flowering from February until April in the forests on the Mufindi escarpment, the Poroto Mts, Bundali Hills and Mbeya range above about 1 800 m. The largest-flowered we have seen were growing by a mountain stream in deep forest above Isoko in the Bundali Hills. A similar large-flowered form is also found in the neighbouring Misuku Hills of northern Malawi.

C.sylvatica is widespread throughout Africa south to Cape Province and also in Madagascar and the Mascarene Islands. It is rather variable particularly in vegetative and floral size and known under a number of later synonyms throughout its range. One of these is *C.stolzii,* a name given to plants collected by Adolph Stolz in the Poroto Mts.

Cynorkis anacamptoides Kraenzl. [58B]
Five species of *Cynorkis* are found in the Southern Highlands, the most widespread of these being *C.anacamptoides.* This small-flowered terrestrial orchid stands 10 to 60 cm tall. Its flower-stalk and flowers are glandular-pubescent and the densely flowered inflorescence may be up to 20 cm long. The flowers range from pink to mauve or purple and are small with the flower segments from 3 and 5 mm long. The lip bears a short pendent spur, 3 to 5 mm long.

It is common in the swamps and in wet grassy areas from 800 to 3 000 m altitude in our area flowering almost throughout the rainy season. It is not uncommon in the marshy valleys beside the road from Mbeya to Tukuyu in the Poroto Mts.

Plate 58. A, *Calanthe sylvatica* (A1, flower); B, *Cynorkis anacamptoides* (B1, flower); C, *C.hanningtonii* (C1, flower)

C.hanningtonii Rolfe and *C.kassneriana* Kraenzl. are also glandular-hairy but both these species are commonly found in montane forest or in plantations in our area.

Cynorkis hanningtonii Rolfe [58C]
This is a showier species than the related *C.anacamptoides* with larger less densely clustered flowers. It reaches 50 cm tall with two basal leaves 2.5 to 13 cm long and 1 to 5 cm broad. The flowers are quite distinctive with a white 5 to 10 mm long, 5-lobed lip, spotted with red. The lip has a slender incurved spur, 3 to 9 mm long, at the base.

It is a common species flowering from March until May frequently being found in the Cypress plantations such as those around Mbeya including those behind the town. These are often rich in orchids which thrive in deep shade which few other plants can tolerate.

Centrostigma occultans (Rchb.f.) Schltr. [IIID & 59A]
Two species, *C.occultans* and *C.clavata,* of this small genus of orchid, related to the ubiquitous *Habenaria,* occur in the Southern Highlands; and of these *C. occultans* is the commoner. It is a terrestrial species varying from 40 to 70 cm tall and the stems bear numerous shortish leaves up to 16 cm long and 1 to 2 cm broad. The green or yellow-green flowers are quite characteristic having a pendent spur, 10 to 15 cm long, and a distinctive lip in which the side lobes are divided into filaments.

In the Southern Highlands *C.occultans* is found around Iringa and in the highlands around Mbeya. A typical locality is a swamp about 10 km past Kisinga mission on the Kitulo road in the Ukinga Highlands. Here it flowers in January and February along with the striking pink-flowered *Gladiolus laxiflorus* and the vivid crimson-flowered *Satyrium robustum.* The recently described iris *Moraea callista* is also found here.

C.clavata is closely related but has a shorter spur with a swollen apex and a lip with longer side lobes up to 4 cm long. It was first discovered in our area as late as 1974 growing in a swamp at Mbozi where it flowers in late February.

Disa fragrans Schltr. (syn. *D.leucostachys* Kraenzl.) [IIIE & 59B]
An aptly-named orchid, 10 to 50 cm tall, with leaves 3 to 17 cm long and 2 cm broad and a cylindrical, dense, many-flowered inflorescence, 3.5 to 12 cm long, 2 to 3.5 cm in diameter. Flowers fragrant, white, usually spotted with pink, or entirely pink, 1 to 1.5 cm across; dorsal sepal 5 to 8 mm long, with a pendent spur 4 to 8.5 mm long.

It is widespread in the grasslands of the Njombe and Mbeya highlands, for example, it flowers in February on the Kitulo Plateau up to 2 800 m and in March on Mbeya Mt towards the summit.

Elsewhere its range extends southwards to Zimbabwe.

Plate 59. A, *Centrostigma occultans;* B, *Disa fragrans;* C, *D.hamatopetala*

Disa hamatopetala Rendle [IIIB & 59C]
Blue-flowered orchids are of great rarity but this species with rich blue flowers and a hanging fringed lip must be one of the most spectacular. Flowering in the dry season from August until October, it often appears after a grassy hillside has been burnt when the long slender leaves are burnt so that the base of the plant resembles a shaving brush from which arises an inflorescence bearing up to 11 flowers. This attractive species, not uncommon in south-western Tanzania, is also widespread in neighbouring Malawi, Zambia, southern Zaire and Zimbabwe, growing on open or shrubby hillsides above 1 500 m. Its close ally *D.longilabris* Schltr., having similarly shaped but smaller purplish flowers, is confined to the Njombe area of the Southern Highlands.

Disa hircicornis Reichb.f. [60A]
A common wet grassland orchid readily recognised by its relatively small dark red-brown, dark purple or rarely pale purple flowers with a hook-like spur on the upper sepal. It is commonly found flowering from December until early March in swamp areas above 1 500 m, often growing by streams. A typical locality is in the swamps by the main Iringa to Mbeya road near Sao Hill at about 2 000 m. *D.hircicornis* is widespread in Africa from Nigeria across to Sudan in the north and to Angola and Zimbabwe in the South. In Kenya and Uganda it flowers in June and July.

Disa robusta N.E.Br. [60B]
A large plant, up to 85 cm tall, with bright orange flowers shaped like a dunce's cap and leaves borne on a separate sterile shoot, the base of the stem often being spotted with purple. It is one of the commonest and most typical orchids of the grasslands up to 2 700 m (but may be found down to 900 m in open areas in woodland) flowering around Christmas-time but still to be seen in flower as late as mid-February. It is called 'vigogwa' by the Wamalila. It is also found commonly in Malawi, Zambia and southern Zaire.

Disa stolzii Schltr. [60C]
A medium-sized but very showy species, up to 60 cm tall, with scarlet flowers; the pink or orange upper sepal spotted with scarlet. It is often to be found flowering in February and March, mostly over 2 000 m altitude, in wetter grassland areas throughout the Southern Highlands. In some localities it grows with the similar but smaller flowered species *Disa erubescens* Rendle (see plate 5A) with which it may sometimes hybridise, for example, near Sao Hill. *D.stolzii* is also wide-spread in Malawi and Zambia.

Disa ukingensis Schltr. [IVD & 60D]
Not unlike *D.engleriana* with which it was until recently mistakenly considered conspecific. It is a shorter plant rarely reaching 35 cm with a denser inflores-

Plate 60. A, *Disa hircicornis* flower; B, *D.robusta;* C, *D.stolzii;* D, *D.ukingensis* flower

cence and shorter broader purple or magenta flowers with a spotted upper sepal. Generally found at higher altitudes in the grasslands above 2 600 m, *D. ukingensis* has so far only been seen on the Kitulo Plateau of south-western Tanzania and on the Nyika Plateau in northern Malawi where it flowers in February and early March.

Disa engleriana Kraenzl., an attractive pink- or purple-flowered orchid, is commonly found in grassland or woodland between 1 750 and 2 100 m. Its slender flowering stem reaches 55 cm and, as with many other *Disa* species, the sterile stem is purple-spotted below. In Tanzania, it is confined to the Southern Highlands where it commonly flowers in February and March. It is also found in neighbouring Malawi, Zambia and southern Zaire.

Disperis anthoceros Rchb.f. [61A]

Despite their small size, the species of *Disperis* are some of the prettiest and often most conspicuous and abundant orchids in the woodlands and especially in the cypress plantations of the Southern Highlands. They have thrived in the latter habitat where the deep shade has eliminated most of the competition met with in their natural environment.

D.anthoceros, one of the commonest of the ten species found in the area, is also one of the most readily recognised. It ranges from 8 to 16 cm tall and in the middle of its stem bears two opposite leaves, up to 3 cm long, often purplish on their lower surfaces. The one to three white flowers borne at the apex of the stem are quite distinctive with their dunce-cap-shaped hood which is 1.5 to 1.7 cm long. The petals are often marked with green veins and the inside of the hood may be marked with pale purple.

The typical variety is common throughout tropical and southern Africa from Nigeria and Ethiopia southwards. In our area it may be seen commonly in the cypress plantations behind Mbeya and also in evergreen and bamboo forests between 1 800 and 2 700 m flowering in February and March.

The large-flowered var. *grandiflora* Verdc. is confined to the Southern Highlands and characteristically has a coarser spur up to 3 cm long. It has been recorded from Imagi Mt near Iringa and from the northern Mbeya range growing on wet logs and rocks by streams in forest between 2 160 and 2 550 m.

Disperis dicerochila Summerh. [61B]

D.dicerochila is amongst the largest flowered species of the genus. It ranges from 6 to 26 cm tall and, like *D.anthoceros,* its stems bear in their middle two opposite leaves up to 5 cm long often purple below. The one to four flowers, borne at the apex of the stem, are attractively coloured white and pink or magenta and are conspicuous, often 2 cm broad across their spreading sepals.

D.dicerochila was first described from Uganda but is widespread in East Africa, Zaire, Malawi and Zimbabwe. In the Southern Highlands it grows in leaf litter in upland rain forest above 1 650 m. However, it is most easily seen

Plate 61. A, *Disperis anthoceros* (A1, lip); B, *D.dicerochila* (B1, lip); C, *D.reichenbachiana* (C1, lip); D, *Epipactis africana*

flowering from January to March in cypress plantations such as that behind the Mbeya hospital or in *Eucalyptus* plantations such as that near Irambo mission in the Poroto Mts.

Disperis reichenbachiana Rchb.f. [61C]
In some localities in the Southern Highlands plants of this delightful orchid are common scattered like jewels on the forest floor. This is particularly true in some of the cypress plantations of the Mbeya and Tukuyu areas, even those near the towns.

D.reichenbachiana ranges from 7 to 25 cm tall. Its small leaves, up to 4.5 cm long, are attractively coloured dark green above with a central silvery stripe and purple below. Commonly it is one- or two-flowered, the flowers lying horizontal to the ground. These are about 1 to 1.2 cm long and attractively coloured rose-pink or pale pink.

This was one of the earliest species described in the genus and it is widespread in tropical Africa from Zaire and Uganda south to Zimbabwe and Angola. However, in Tanzania, it is confined to the Southern Highlands where it is particularly common in many areas from 1 350-2 100 m altitude. Apart from the Mbeya plantations, it is common around Mufindi and also around Tukuyu both in natural forests amongst the leaf litter, in plantations and also in secondary grassland and bushland. It flowers in February and March.

Epipactis africana Rendle [61D]
The tallest of all African terrestrial orchids is undoubtedly *E.africana,* one of only three African representatives of a predominantly European genus. Its reddish-hairy slender stems can reach 3.5 m tall with, along their length, large recurved leaves up to 20 cm long. The leaves are smaller towards the apex merging eventually into the bracts which subtend the 2 to 2.5 cm long flowers. The flowers are greyish-purple on the outside and greenish-yellow within, veined with purple on the petals and with a yellow-brown lip.

It usually grows in montane evergreen and bamboo forest and also in bush, bracken or coarse grass by rivers about 2 000 m. It is not uncommon in the forests of the Poroto Mts and on Mt Rungwe where it can be found flowering in October. The Wamalila call it 'ndungulingu'.

E.africana is widespread in Africa from Ethiopia south to Malawi.

Eulophia ischna Summerh. [62A]
This delicate terrestrial orchid, although common in south-central Africa, is relatively rare in the Southern Highlands.

It is a slender erect plant, up to 45 cm tall, with two slender grasslike leaves, up to 26 cm long and 3 mm broad, produced from a white fleshy subterranean tuber. The inflorescence of several well-spaced drooping flowers is produced at the same time as the leaves. Flowers, 1.5 to 1.8 cm long, white, tinged with red

Plate 62. A, *Eulophia ischna;* B, *E.ovalis;* C, *E.orthoplectra* flower; D, *E.shupangae;* E, *E. streptopetala*

especially at the base of the segments; lip purple, with a yellowish base, with a short purple spur, 2 mm in length, at its base.

We have found this plant in some quantity around Sao Hill in marshes at 2 100 m by the main Iringa to Mbeya road flowering in January. Mary Richards also found it between Mwakete and Njombe in similar habitats.

Many other species of *Eulophia* are found in the Southern Highlands. Most are still rather poorly known but *E.cucullata* (Sw.) Steud. is frequently seen in the wet grasslands at medium altitudes. It is a large plant, often a metre tall, with large pink or rose-purple flowers having a two-lobed callus on the saccate lip. The largest species of all is, however, *E.horsfallii* (Batem.) Summerh. found by the margins of streams and in wet glades in riverine forest. It often grows to 2 m tall and has several large maroon and purple flowers with a yellow crest on the lip. We have found it growing by a stream at Chivanjee.

Eulophia ovalis Lindl. [62B]

Another *Eulophia* species whose presence is seldom detected until it flowers because it has rather grasslike leaves well hidden by its typical grassy habitat.

The five or so spreading leaves, up to 30 cm long and 0.5 to 2 cm broad, are produced from a subterranean fleshy tuber. Inflorescence, commonly about 25 to 30 cm long, of a few flowers about 4 cm across. The canary yellow petals lie almost horizontal and are the showiest part of the flower. Sepals usually pale green or brownish green; lip also yellow, spotted with brown, with a yellow callus and a pinkish spur, 4 mm long, at its base.

Nowhere common but can often be seen growing in ones and twos in most of the moorlands of the area. We have seen several plants in January on the Kitulo Plateau on the slopes of Ishinga Mt at about 2 700 m and, in February, on Mbogo Mt, Umalila at 2 500 m. It has also been collected on Lukota Mt in Iringa District at about 2 000 m and on Kito Mt, Sumbawanga at 1 900 m.

Elsewhere it is a widespread but local species south to South Africa.

Eulophia shupangae (Rchb.f.) Kraenzl. [62D]

This, probably the commonest species of *Eulophia* in the Southern Highlands, is a terrestrial orchid, up to 75 cm tall, with two grasslike leaves, up to 50 cm long and 1 cm broad, produced from a white underground tuber. Inflorescence axis characteristically covered by pale brown scarious sheaths with a cluster of flowers subtended by prominent bracts. Flowers are about 1.5 cm in diameter with a 1 cm long lip and a 3 mm long spur at its base. Two colour forms are found almost throughout its range, the more obvious having bright yellow flowers with red papillae on the lip whilst the drabber form has pale mauve or maroon flowers.

The colour variants are often found growing together in the same area and both can be seen by the roadside and in the grasslands beside the Mbeya to

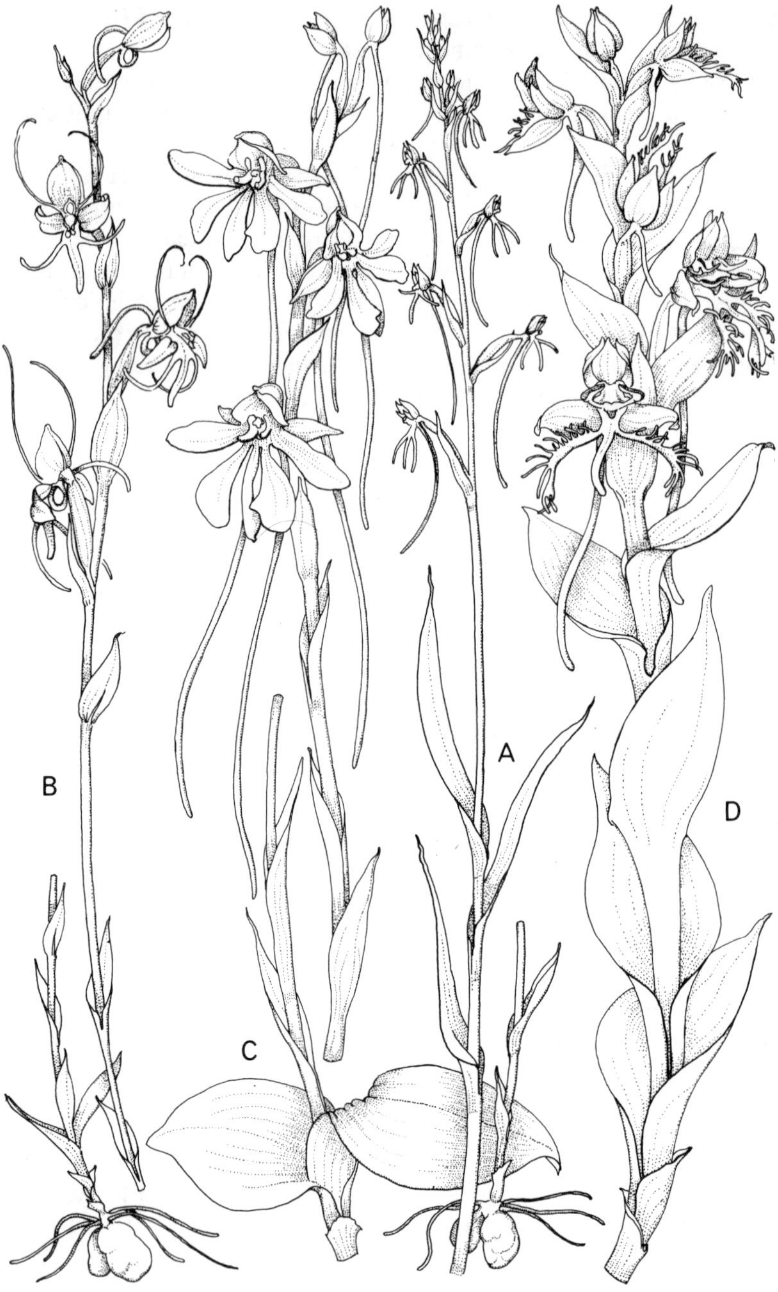

Plate 63. A, *Habenaria filicornis;* B, *H.goetzeana;* C, *H.macrura;* D, *H.occlusa*

Chunya road over 1 800 m. We have also seen them in the Poroto Mts and also around Sao Hill at 2 000 m flowering from January until March.

E.shupangae is widespread throughout tropical Africa.

Eulophia streptopetala Lindl. [62E]

This attractive terrestrial or lithophytic orchid, better-known in East Africa as *E.paivaeana* (Rchb.f.) Summerh. (a later name), is a widespread species ranging from the Yemen and Ethiopia to South Africa.

A stout plant, 50 to 100 cm tall, with ovoid green pseudobulbs and stout white roots. About five pleated leaves, up to 60 cm long and 8 cm broad, are produced as the plant begins to flower on the upper part of a long erect unbranched several-flowered inflorescence. Flowers, 3 to 4 cm across, with bright canary yellow petals, a yellow lip marked on the sides with red, and green sepals mottled with brown or purple and a short red spur, 4 to 6 mm long, at its base.

This orchid is frequently seen in the Southern Highlands but is only obvious when in flower between November and January. It grows in grassland on the Kitulo Plateau at 2 600 m and in dry meadows and on the margins of thickets around Sao Hill at 2 000 m. In Ufipa, it is also common and can be seen at 1 850 m on the Malonje Plateau on the road to Mpui and in hollows in open grassland near Sumbawanga at 2 200 m.

E.orthoplectra (Rchb.f.) Summerh. [62C]

This species is closely allied to *E.streptopetala* and almost equally widespread. It has large yellow and red flowers with a longer conical spur but flowers from May until August in the dry woodlands of Ulambya and Mufindi.

Habenaria filicornis Lindl. [63A]

Over 50 species of *Habenaria* occur in the Southern Highlands, nearly one half of all those found in East Africa. However, they are never as striking an element in the flora of the grasslands or forests as are the colourful species of *Disa* and *Satyrium*. In fact, most of the species of *Habenaria* are rather drably coloured in comparison, having predominantly green or white flowers. Even so, some of the species are attractive plants and, in all, the intricate nature of their flowers make them a fascinating group to study.

One of the smallest flowered species is the common *Habenaria filicornis,* a plant, up to 50 cm tall, with slender leaves 3 to 11 cm long. The 5- to 20-flowered inflorescence is slender and each green flower has a 3-lobed lip about a centimetre long with a slender incurved spur, 1.5 to 3 cm long, at the base. *H. chlorotica* also found in the grassland of our area is closely allied but differs in having longer leaves and smooth rather than papillose stigma lobes.

In our area *H.filicornis* is common in woodlands, in secondary bushland and also in grassland from 1 350 to 1 950 m, flowering in late December and January.

Plate 64. A, *Habenaria adolphii;* B, *H.cornuta;* C, *H.malacophylla;* D, *H.welwitschii*

Habenaria goetzeana Kraenzl. [63B]

Extraordinary flowers are common among the many species of *Habenaria* but, in this species, the combination of unusual shape with a delicacy of form is remarkable.

H.goetzeana was named after Goetze one of the earliest plant collectors in the area who indeed lost his life on the northern shores of Lake Nyasa and is buried at Lumbila. It is a small orchid up to 35 cm tall with several small leaves 2 to 3.5 cm long. The pale green flowers are large for the plant, about 1.5 cm across, with the 2 to 3 cm long spur looping over the ovary behind the flower, with its apex finally pointing backwards.

It is a common species in our area found growing in the woodlands and lower grassland areas from 950 to 2 100 m altitude and flowering in February and March. A typical locality is in the grassland by the village of Isongole in the Poroto Mts where it grows with *Satyrium* species and *Neobolusia stolzii* amongst scattered *Clematopsis* plants. It is also found elsewhere in western and southern Tanzania and in neighbouring Zambia and Zaire.

Habenaria macrura Kraenzl. [63C]

This lovely white-flowered orchid may be seen in its thousands on the high moorlands of the Kitulo Plateau in early February. One of the best spots to see it is on the road to the Kitulo sheep project farm some 5 km beyond Kikondo where the road skirts the cliffs of Chaluhangi Mt.

It is one of a group of *Habenaria* species which have two leaves which are appressed on the soil surface at the base of the plant. The leaves are 2.5 to 9.5 cm long, 2.5 to 7 cm broad and from between these emerges the inflorescence up to 65 cm long. The flowers are relatively large with a 9 to 17 cm long pendent spur emerging from the base of the lip and often hidden by the bracts.

H.macrura is widespread in Africa from Nigeria, Ethiopia and Uganda south to Zambia and Malawi. Although most frequently seen in grassland, it is also not uncommon in woodlands from 1 200 to 2 400 m.

Habenaria occlusa Summerh. [IIIC & 63D]

This large-flowered orchid is only found in the Southern Highlands and was described as recently as 1964 from a plant collected by Mary Richards on the Kitulo Plateau.

It is one of the largest flowered *Habenaria* species belonging to a group characterised by the comb-like side lobes of the lip. The plant reaches 40 cm tall with stem-leaves up to 8 cm long. The yellowish green or pale green flowers are about 3 cm across with a somewhat S-shaped 1.5 cm long spur at the base of the lip.

So far, *H.occlusa* has been seen only on the Kitulo Plateau and on Mbeya Mt.

It is closely allied to the larger *H.praestans* and *H.splendens* but differs in its denser inflorescence, smaller flowers and shorter spur. Both *H.praestans* and *H.splendens* are common in our area but are generally found growing at lower altitudes than *H.occlusa.*

Habenaria adolphii Schltr. [64A]
An attractive terrestrial orchid, 30 to 50 cm tall, growing from an ellipsoid tomentose tuber. It has two basal fleshy leaves, 5 to 7 cm long and 5 to 8 cm broad, strongly appressed to the ground and several smaller leaves along the stem. Inflorescence, 4 to 12 cm long, with 3 to 15 white or creamy-white flowers about 3 to 3.5 cm long with green sepals and a 2 to 2.5 cm long spur at the base of the lip. The ovary is quite characteristic with six longitudinal toothed wings.

We have found this species in flower in February at Songwe, west of Mbeya, where it grows in wooded grassland on limestone, a habitat more famous for its rich asclepiad flora. It has also been collected on the Muse to Kisungu road in Ufipa at 1 500 m.

H.adolphii has also been recorded from southern Zaire.

Habenaria cornuta Lindl. [64B]
A very variable terrestrial orchid, 20 to 80 cm tall, growing from an ellipsoid or ovoid tuber. The leaves, up to 10 cm long and 4.5 cm broad, are glabrous. The inflorescence is laxly four- to many-flowered and up to 19 cm long. The pale green or yellow-green flowers are spidery, 1.5 to 3 cm long with a 1.4 to 2.7 cm long spur at the base of the lip.

It is widespread in Africa from Ethiopia and Nigeria south to South Africa and is commonly found growing in poorly drained grassland and in *Brachystegia* woodland up to about 2 400 m. In the Southern Highlands it is frequently found in flower from January until March. It grows in the marshes around Sao Hill at about 2 000 m and in the woodlands of the Chimala escarpment east of Mbeya.

Habenaria malacophylla Rchb.f. [64C]
A terrestrial orchid, 30 to 100 cm tall, with a tomentose, ovoid or ellipsoid tuber and thin-textured leaves, 7 to 20 cm long, the longest being in the middle of the stem. The slender inforescence, 8 to 34 cm long, is laxly many-flowered. The flowers are small, rather spidery and green. The dorsal sepal is about 4 to

PLATE IV

A	B
C	D

A *Disa erubescens,* near Sao Hill, Mufindi, p. 203
B *Calanthe sylvatica,* near Mwakaleli, Livingston Mts, p. 199
C *Satyrium chlorocorys* (left), p. 223; and *Satyrium acutirostrum,* p. 222, Kitulo Plateau
D *Disa ukingensis,* Kitulo Plateau, p. 203

6 mm long and the lip, 4.5 to 7 mm long, has a slender spur, 9 to 18 mm long, at its base.

Essentially, it is a species of upland rain-forest to about 2 400 m, but occasionally it can be found in upland grassland. We have found it not uncommonly in the Poroto Mts, for example, it grows in bamboo thickets on Ngosi Mt along with *Calanthe sylvatica.*

Elsewhere, it is throughout tropical and southern Africa.

Habenaria welwitschii Rchbf. [64D]

A terrestrial orchid, 20 to 70 cm tall, growing from an ellipsoid sparsely tomentose tuber with leaves 5 to 26 cm long and 1 to 1.5 cm broad. The inflorescence, 6 to 18 cm long, is rather loosely 10- to 30-flowered. The flowers resemble a greenish-white gnat or mosquito with the segments less than 1 cm long but with a swollen club-shaped spur, 1 to 1.6 cm long, at the base of the lip.

It is particularly common in the swamps and marshes around Sao Hill and Mufindi at about 1 900 m where it often grows in shallow water, flowering in January and early February. It is also to be seen around Mbozi in a similar habitat but at a slightly lower altitude of 1 500 m.

Elsewhere it is found in western Uganda, Rwanda, Mozambique and Angola.

Several closely related species, all with mosquito-like flowers, are found in the Southern Highlands. The identification of these is difficult and almost impossible without recourse to Summerhayes' excellent account of the genus in the *Flora of Tropical East Africa.*

Holothrix hydra Cribb [65A]

Most species of the terrestrial orchid genus *Holothrix* are extremely hairy. Eight species of the 15 found in East Africa are found in the Southern Highlands. Of these, *H.hydra* could justly lay claim to being the most attractive, indeed it is one of the most beautiful and delicate of all the orchids found in the area.

From between two basal prostrate hairy leaves, about 5 to 6 cm long, arises a slender pubescent inflorescence, up to 50 cm long, at the apex of which are borne a few pure white flowers with highly dissected petals and a similar lip with a short incurled spur at its base. The petals and lip are up to about 3 cm long.

H.hydra is found only in the Mbeya and Ufipa highlands. We have found it in flower in February growing commonly in colonies on the slopes in open woodland on the escarpment above Chimala at about 1 800 m.

It is closely allied to *H.longiflora* Rolfe, a species widespread in South-central Africa. The latter differs however in its almost glabrous leaves and flowers in which the divisions of the petals and lips are much more slender and filiform.

Plate 65. A, *Holothrix hydra;* B, *H.nyasae;* C, *H.puberula;* D, *Neobolusia stolzii*

Holothrix nyasae Rolfe [65B]

H.nyasae must be one of the hairiest of all orchids and, although growing up to 40 cm tall, its tiny flowers render it relatively inconspicuous.

It has two prostrate basal leaves, up to 4 cm long and 6.5 cm broad, covered with long rather soft hairs. The slender inflorescence axis is also covered with long spreading hairs as are the sepals. The white or yellowish flowers, 5 to 9 mm long, all face to one side in a slender elongated raceme. The petals are 3-lobed, but the fan-shaped lip may be 7- to 13-lobed and has an incurved conical spur, about 2 mm long, at its base.

It is apparently confined to the mountains around the top end of Lake Nyasa where it grows amongst rocks and in short grassland between 2 000 and 2 950 m. We found it in early February on the very summit of Mtorwi Mt, the highest point in the Southern Highlands, and on the edge of the Kitulo Plateau. There, in a natural rock garden, *H.nyasae* and many other orchids grow with the beautiful blue and white iris *Moraea callista* and several species of bushy heath. It is also found less commonly on Mbeya Mt, the Umalila plateau and the Bundali hills.

Holothrix puberula Rendle [65C]

Whereas most species of *Holothrix* have white flowers and a divided lip, *H. puberula* differs markedly for it has a distinctive entire blue or violet lip which is much longer than the other floral segments.

It is a tiny plant seldom reaching 20 cm tall, the hairy inflorescence bearing between one and five flowers, each 7 to 10 mm long. The almost glabrous flowers have an oblanceolate or elliptic lip with a narrowly conical, incurved, 3.5 to 4 mm long spur at its base.

This pretty orchid is not uncommon in the mountains around Mbeya but is difficult to find in its natural habitat of mountain grassland because of its small size. However, it is relatively easily found growing with other orchids such as *Disperis, Nervilia, Malaxis* and *Cynorkis* species in the cypress plantations around Mbeya. Some of the richest of these are behind the General Hospital in Mbeya itself where it can be seen in flower in late January and February.

Elsewhere in East Africa, *H.puberula* is widespread as far north as Ethiopia but is at the southernmost extent of its range in the Southern Highlands.

Neobolusia stolzii Schltr. [65D]

Few East African orchids are as easy to recognise as *N.stolzii,* a small terrestrial species, up to 45 cm tall, with several small bee-like flowers.

It has 2 to 4 leaves, up to 9 cm long and 1.2 cm broad. The flowers, which lack a spur, are about 1 cm across and have yellowish sepals and petals and a convex, brownish maroon, furry lip with a delicate white marginal fringe.

N.stolzii is the northernmost representative of a small genus of predominantly South African orchids which reaches its northern limit in the Mbeya

Plate 66. A, *Liparis bowkeri;* B, *L.nervosa* (B1, flower); C, *Malaxis weberbaueriana* (C1, flower); D, *Nervilia reniformis*

region. Despite that, it is not uncommon in the area and may be seen flowering in January and February in the grasslands of the Poroto Mts near Isongole and, again, on the slopes of Mt Rungwe between 1 850 and 2 600 m.

Var. *bombyliflora* Cribb which has a larger lip lacking the marginal white fringe has been collected infrequently in the Poroto Mts e.g. in the grassland around the village of Isongole.

N.stolzii is also found in neighbouring Malawi and in Eastern Zimbabwe.

Liparis bowkeri Harv. (syn. *L.neglecta* Schltr.) [66A]
This is one of the commonest terrestrial orchids in the area. It is often a small plant but can reach 30 cm tall in favourable localities. The pseudobulbs are cylindric to narrowly conical, up to 7 cm long, and are characteristically borne two together at an angle to each other, the new one being erect. The three or four pleated leaves are 6 to 12 cm long and 3 to 6 cm broad. The inflorescence reaches 15 cm long and bears a few greenish to ochre-coloured flowers about 7 to 9 mm across.

The easiest places to find *L.bowkeri* are the cypress plantations which are widespread around Mbeya and Tukuyu. Here, *L.bowkeri* grows in patches with *Malaxis weberbaueriana, Nervilia* and *Disperis* species. However, its natural habitat is in the woodlands and montane forest patches of the area where it commonly flowers in January and February. It is a widespread species elsewhere in Africa from Uganda south to South Africa.

Liparis nervosa (Thunb.) Lindl. [66B]
L.nervosa, the largest species of *Liparis* found in East Africa reaching 45 cm tall, has fleshy basal whitish or greenish pseudobulbs, up to 1.6 cm in diameter, hidden by the leaf bases. The pleated leaves reach 16 cm long and 6 cm broad and the slender inflorescence has a ridged stalk. The flowers, green or yellow with a maroon lip, are small, about 5 mm across.

This species is one of the most widespread of all orchids being virtually pantropical. It is widespread in tropical Africa where it has been known under various names, the more notable *L.rufina* Lindl. and *L.guineensis* Lindl., both of which are later synonyms.

In the Southern Highlands it grows in a variety of habitats, in woodlands, forests and plantations and also in wet grassland where it generally has shorter thicker more obviously pleated leaves. It commonly flowers from December until March in our area.

Malaxis weberbaueriana (Kraenzl.) Summerh. (syn. *M.stolzii* Schltr.) [66C]
Although a species of wet woodland and montane forest, this attractive terrestrial orchid is most frequently seen growing in small groups on the deeply shaded floors of the numerous cypress plantations around Mbeya and Tukuyu.

It is a delicate herb, up to 25 cm tall, growing from a fleshy rhizome. The 3

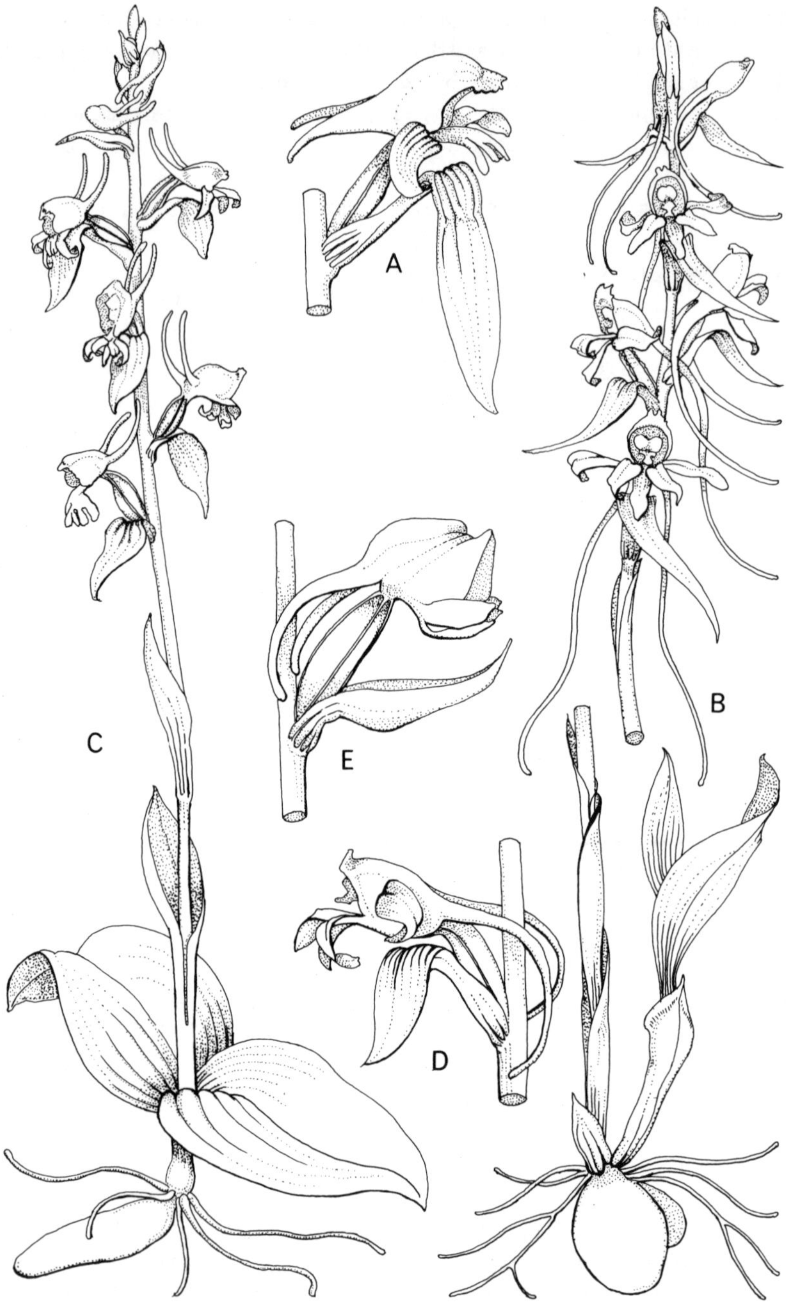

Plate 67. A, *Satyrium atherstonei* flower; B, *S.buchananii;* C, *S.comptum;* D, *S.neglectum* flower; E, *S.robustum* flower

to 4 pleated leaves are borne at the apex of a short fleshy stem and are each up to 5.5 cm long and 3 cm broad. The tiny green or purple flowers, each 4 to 5 mm across, are borne on a slender stalk up to 17 cm long.

In our area, it grows from 900 to 2 100 m, a typical locality being in the plantation at Mpuguso just south of Tukuyu where it flowers from January until March. Elsewhere it is widespread in tropical Africa from Kenya south to Zambia.

Nervilia reniformis Schltr. [66D]

Among the most difficult of all orchids to identify are the species of *Nervilia.* The reason for this is twofold for, firstly, several species can grow together in one place and, secondly, the flowers appear and set seed before the solitary leaf appears and the leaves in general are more characteristic of the species than the flowers.

Nervilia species fall into two categories, those with two or more flowers per stalk and those with one flower only. *N.reniformis* is the commonest of the one-flowered species and easily recognised by its prostrate reniform leaf which is hairy on its upper surface. The upward pointing flower is about 1.5 cm long and has a finely pubescent and papillose lip with a ragged apical margin. The flowers have brownish-green sepals and petals and a white lip with a yellow centre.

It is widespread in the woodlands and less commonly in grassland in the hills north of Lake Nyasa. In the valleys leading down to the Kiwira River such as that on the Chivanjee Tea Estate, *N.reniformis* is commonly found growing in open woodland with several other species such as *N.kotschyi* (Rchb.f.) Schltr. with an erect ovate glabrous leaf with a purple underside, and *N.adolphii* Schltr. with a smaller spreading glabrous somewhat angular leaf. It flowers in November and December and the leaves appear in January and February.

Satyrium atherstonei Rchb.f. [67A]

One of the commonest and most distinctive orchids of the area, readily identified by its hairy ovary and small short-spurred white flowers almost hidden by reflexed and predominantly white bracts. It is common in grassland from 900 to 1 900 m altitude often by streams or swampy areas, for example, at Kibenga in the Dabaga Highlands and near Sao Hill in Mufindi. *S.atherstonei* is widespread throughout West and Central Africa from Guinée and Uganda in the north as far south as the Transvaal and Cape Province of South Africa.

Satyrium buchananii Schltr. [67B]

A widespread and attractive terrestrial orchid, 25 to 60 cm tall. Leaves 3 to 5, borne on a sterile stem, the lower sheathing, the upper lanceolate, 7 to 20 cm long and 1.5 to 3.7 cm broad. Inflorescence 6 to 15 cm long, rather laxly 7- to

20-flowered. Flowers white, fragrant; lip hooded, 10 to 15 mm long, with two characteristic slender spurs 4 to 7 cm long.

It is widespread and common in the seasonally wet grasslands at intermediate altitudes (1 000 to 2 400 m) in our area. The Wakinga in the mountains west of Njombe call it 'lisekeni' and eat its tubers in times of famine. Near Ibaba, in the western Bundali Hills at 2 100 m, hundreds of plants produce a carpet of white on the mountainside in early February and it is almost as common in the marshes around Sao Hill in January.

In our area, it reaches the northern limit of its distribution which extends south to eastern Zimbabwe.

On the stony hillsides and in rocky places at higher altitudes, this species is replaced by the similar *S.longicauda* Lindl. distinguished by its shorter spurs, 2.5 to 4.5 cm long.

Satyrium comptum Summerh. [67C]

As rare as *S.aberrans,* and often found growing with it amongst rocks at or near the summit of the highest peaks at the western end of the Southern Highlands. It is an attractive dwarf species, 7 to 20 cm tall, with several distinctive white flowers with 1 cm long spurs at the back of the flower curving away from the ovary. So far, it has only been found on Rungwe Mt, Mbeya Peak and on Mtorwi Mt, the highest of the many peaks which surround the Kitulo Plateau.

Satyrium neglectum Schltr. [67D]

A ground orchid, 20 to 100 cm tall, with developed leaves, 12 to 27 cm long, borne on a separate sterile shoot. Inflorescence dense, many-flowered, 5 to 25 cm long. Flowers pink or, more rarely, white or creamy white; lip hooded, 5 to 9.5 mm long, with two parallel slender spurs 6 to 17 mm long.

Two varieties are known in our area: var. *brevicalcar* Summerh. recorded from Mdapo near Njombe is the rarer with spurs less than 10.5 mm long; the typical variety with spurs over 13.5 mm long is widespread and can be seen, for example, in thousands on the Kitulo Plateau, in the Porotos and on Mt Rungwe in mid-February.

Elsewhere it is widespread from Mt Kilimanjaro south to the Cape. The allied species *Satyrium acutirostrum* Summerh. [Plate 5C] with its robuster spikes of larger, almost globose, orange-red flowers is common from January until March in the grasslands of the Mbeya Range, Poroto Mts, Kitulo Plateau and Mt Rungwe. Elsewhere its range extends from southern Zaire to Zimbabwe.

Satyrium robustum Schltr. [67E]

A spectacular orchid, 25 to 70 cm tall, with several mostly basal leaves up to 28 cm long and 5.5 cm broad and a cylindrical dense 11- to many-flowered inflorescence up to 24 cm long. Flowers bright pink to deep crimson; lip hooded, 7 to 10 mm long, with slender parallel spurs 12 to 18 mm long.

We have found it growing in late January at 2 200 m near Kisinga on the Kitulo Plateau in a swamp growing with another orchid *Centrostigma occultans* and the tall pink *Gladiolus laxiflorus.* Elsewhere on the plateau, it grows around the Kitulo range, by the Ndumbi River and in other swampy areas but is always rather localised. The same is true elsewhere in the Southern Highlands but it has been recorded in October and November from Iheme near Iringa, Njombe and the Kiwira Fishing Camp on Mt Rungwe.

It is also found on eastern Uganda and in Kenya.

Satyrium aberrans Schltr. [68A]

A tiny and unmistakable orchid, up to about 17 cm long, with 3 to 5 stem-leaves, 2 to 6 cm long, and a dense few- to many-flowered inflorescence 2 to 4 cm long. Flowers white, subtended by chocolate-brown bracts; lip hooded, 3 to 4 mm long, with two cylindrical spurs 3.5 to 6 mm long.

This high mountain species prefers rocky cliffs and is known only from the Njombe and Mbeya highlands and from eastern Zimbabwe. Typical localities in our area are on the summits of Mtorwi and Mbeya Mts. In the latter locality, it grows with *Commelina hockii, Aristaea alata, Cephalaria pungens, Thalictrum zernyi* and the allied *Satyrium comptum.*

Satyrium chlorocorys Rolfe [IV C & 68B]

A terrestrial herb, 20 to 90 cm tall, with slender to stout erect stems and several leaves, the middle ones spreading and up to 24 cm long and 9 cm broad, the rest sheathing or adpressed. Inflorescence cylindrical, up to 37 cm long, rather dense, many-flowered. Flowers green, yellowish green or rarely whitish; the hooded lip 5 to 6.5 mm long and the slender parallel spurs 8 to 22 mm long.

Widespread in the higher montane grasslands and rocky places of the area, flowering from late January until March. It can be seen above Mbeya town and throughout the Mbeya range in suitable localities, on the Kitulo Plateau and in the Umalila and Njombe highlands above about 1 800 m altitude.

Elsewhere, it is found on Kilimanjaro and Mt Meru in northern Tanzania and around Songea in the south, and also in Uganda, Malawi, Mozambique and eastern Zimbabwe.

Schizochilus sulphureus Schltr. [68C]

This attractive terrestrial orchid is the most northernly representative of a small genus of predominantly South African orchids.

It is one of the most distinctive terrestrial orchids of the area. Although it is a small plant, at the most 18 cm tall, with several small leaves up to 7 cm long, it is one of the few species with bright yellow flowers. These are small, mostly 4 to 6 mm long, borne clustered in an arching raceme.

It is fairly widespread on the mountains of the Mbeya region where it commonly grows in grassland or on rocky slopes from 2 000 m to 2 900 m. It may

Plate 68. A, *Satyrium aberrans* (A1, flower); B, *S.chlorocorys;* C, *Schizochilus sulphureus* (C1, flower); D, *Stenoglottis fimbriata*

be seen in flower on Mbeya Peak, on the Kitulo Plateau and in the Bundali Mts in flower from January until March.

S.sulphureus is also found in neighbouring Zambia and Malawi but nowhere else in East Africa.

Stenoglottis fimbriata Lindl. [68D]

This is one of the few African orchids which are fairly easily grown in cultivation. It is a delightful orchid growing up to 35 cm tall from a basal rosette of undulate leaves, up to 12 cm long, which are often spotted with purple.

The inflorescence consists of several pretty pink to pale mauve flowers spotted with purple on the lip and all facing one way. The lip is 7 to 9 mm long and extends beyond the other floral segments, all of which are borne at the apex of a papillose ovary.

Like *Schizochilus sulphureus* and *Neobolusia stolzii* this species is the most northernly representative of its genus. Although strictly speaking a terrestrial orchid, *S.fimbriata* is a forest species and commonly grows on the mossy boughs or trunks of living and fallen trees. It flowers from March until May in the mountains around Mbeya and Njombe. On one tree by the Kiwira to Ndaga road in Unyakyusa this orchid was found growing with three other species, *Bulbophyllum stolzii, Polystachya malilaensis* and *Chamaeangis sarcophylla.* It is also found around the Ngosi Crater Lake and in the forests around Isoko in the Bundali Mts.

Platycoryne buchananiana (Kraenzl.) Rolfe [69A]

A pretty terrestrial orchid, 20 to 65 cm tall, growing from an ellipsoidal tuber. The leaves are 2.5 to 9 cm long and 5 to 13 mm broad. Its clustered flowers are commonly bright orange or more rarely yellow-orange and 1 to 1.5 cm long. The spur, 1 to 1.8 cm long, is at the base of the entire ligulate lip.

It grows in our area in damp or marshy grasslands or in shallow soil over rocks up to about 1 850 m and flowers in January and February. Elsewhere it is found in neighbouring parts of Malawi, Zambia and Zaire and also in Zimbabwe.

P.protearum (Rchb.f.) Rolfe, a closely allied species, is also found in our area but can be distinguished by its laxer inflorescence and its rostellum which has shoulder-like rather than slender side-lobes.

Pterygodium ukingensis Schltr. [69B]

An erect terrestrial orchid, up to 20 cm tall, growing from an ovoid tuber. The lower leaves reach 11 cm long and 3.5 cm broad, the upper ones being smaller. The rather densely 7- to 11-flowered inflorescence is 4 to 6.5 cm long. The clear-green flowers are about 8 to 10 mm in diameter.

This is the most northerly representative of a predominantly South African

Plate 69. A, *Platycoryne buchananiana;* B, *Pterygodium ukingensis;* C, *Roeperocharis bennettiana;* D, *Schwartzkopffia lastii*

genus of orchids. It is found only on the Kitulo Plateau where it grows in the montane grasslands around Kitulo and on the slopes of the fringing mountains.

Roeperocharis bennettiana Rchb.f. [69C]

A slender terrestrial orchid, 35 to 95 cm tall, growing from an ellipsoid underground tuber. The leaves are up to 25 cm long and 2.5 cm broad. The narrowly cylindrical inflorescence is 9 to 27 cm long with yellow-green or green flowers about 1.5 to 2 cm long. They resemble rather those of *Habenaria* but the column is quite distinct.

Roeperocharis is a small genus of only two species both of which occur in our area. The one illustrated here, *R.bennettiana,* is not uncommon on Mt Rungwe and in the grasslands of the Kitulo Plateau over about 2 400 m. We have found it in flower in February growing in bare patches between clumps of grass in the valley below the Kitulo Farm Project buildings. Elsewhere it is found in Ethiopia and Kenya and in neighbouring Malawi.

The other species, *R.wentzeliana* Kraenzl., is similar to *R.bennettiana* but differs in the broader undulate petals and shorter spur up to 7.5 mm long. It has been collected in May at Mufindi growing in marshy ground.

Schwartzkopffia lastii (Rolfe) Schltr. [69D]

A terrestrial leafless pink orchid, up to 11 cm tall, with thick fleshy roots. One to three flowers are produced at the apex of the stem with pink and white petals, 1 to 1.6 cm long, and a mauve lip, up to 2 cm long, with a central yellow spot.

This is a rarely found saprophyte of *Brachystegia* woodland and wooded grasslands up to about 1 500 m. We have found it once or twice in the woodlands at the north end of Lake Nyasa around Kyimbila and Tukuyu where it flowers in December and January.

Elsewhere it is recorded from throughout south-central Africa.

Aerangis carnea J.Stewart [70A]

Only described recently, this orchid has been collected fewer than half a dozen times since its discovery nearly 50 years ago by Miss Cunningham.

It is a small epiphyte found growing in the shade on trees in forests and often over small streams in the mountains. Its leaves reach 15 cm long and one or two arching or hanging 2- to 7-flowered inflorescences, up to 8 cm long, are borne on each plant. The star-shaped flowers are quite large, from 5 to 5.5 cm across, and are white or pinkish white each with a 5.5 to 6.5 cm long pink spur.

It is allied to *A.kirkii* a species widespread in lowland Tanzania growing between sea-level and 450 m but it has a longer column and petals as broad as the sepals.

A.carnea seems to be very localised having been found so far only in the Poroto Mts between 1 450 and 1 850 m altitude, in Ulambya at 1 500 m and at

Plate 70. A, *Aerangis carnea;* B, *A.montana;* C, *A.oligantha;* D, *Angraecum stolzii*

Kyimbila in Unyakyusa at 1 400 m. We have found it growing over a stream in riverine forest at Ikuyu, a few kilometres north of Irambo mission.

Aerangis montana J.Stewart [70B]

A.montana is a well named orchid growing as it does from about 1 800 to 2 400 m altitude. It was described in 1979 based on a plant collected at Igonat near Njombe.

It is a robust epiphytic plant with a very short stem, up to about 6 cm long, bearing 2 to 6 leaves each from 10 to 16 cm long. The arching or pendent inflorescences bear 4 to 10 or more flowers and reach 30 cm long. The flowers are 2 to 2.5 cm across and open pale green but turn white. Each is borne on a salmon-pink stalk and have a pink spur, 5 to 6 cm long.

A.montana grows in montane forest low down on the branches of trees such as *Ilex mitis* with its roots covered by mosses and lichens. It is not uncommon in the Southern Highlands from Imagi Mt near Iringa south-west to the Poroto Mts and on Rungwe Mt where it commonly flowers from November to December. Its distribution extends to adjacent parts of Malawi and Zambia. It has frequently been confused in the past with the smaller flowered *A.mystacidii* (Rchb.f.) Schltr. which is common in south-central Africa and has also been found in the Mbaka valley in Unyakyusa.

Aerangis oligantha Schltr. [70C]

This is a pretty little species with white flowers which is widespread throughout the Southern Highlands, being present in most forests over 1 600 m altitude, but has so far not been found elsewhere.

It has short stems, up to 7 cm tall, bearing 4 to 8 leaves, up to 8 cm long. The short inflorescences reach 4 cm long and bear up to 12 flowers rather clustered together. The flowers reach 1.6 cm across and open pale green before turning white. Each flower has a short 8 to 13 mm long clavate spur.

Like *A.mystacidii* Schltr. and *A.appendiculata* (De Wild.) Schltr., both from south-central Africa, it has a short lip but differs in its much smaller short-spurred clustered flowers.

In the forests of the Mbeya, Njombe and Iringa highlands at altitudes between 1 000 and 2 100 m, one can find *A.oligantha* growing on low bushes and on small trees with its roots covered by mosses and lichens. It flowers in late October into November.

Angraecum stolzii Schltr. [70D]

Several species of *Angraecum* are to be found in Tanzania most being small epiphytic plants of woodlands and forests.

Angraecum stolzii is widespread in the montane forests of the Southern Highlands. It has elongated, slightly compressed, erect or more commonly pendulous leafy stems, up to 30 cm long, often growing in dense masses. The leaves are

Plate 71. A, *Bolusiella iridifolia* subsp. *picea;* B, *Bulbophyllum scaberulum;* C, *B.stolzii;* D, *Cyrtorchis crassifolia;* E, *Diaphananthe stolzii*

borne distichously and are mostly less than 8 cm long. The inflorescences are commonly one- or rarely two-flowered and are borne at each node in bunches of two or three. The small green flowers turn buff as they mature and are 1 to 1.5 cm across and with a short spur less than 1 cm long.

It flowers in November and curiously again towards the end of the rains in April and it is frequently met with in the montane forests of Mufindi and the Mbeya region. It is also common in Malawi.

Less frequently seen, probably because of their small size, are two other small *Angraecum* species with short stems and minute flowers. These are *A.sacciferum* Lindl. which flowers from February until April and *A.chamaeanthus* Summerh.

Bolusiella iridifolia (Rolfe) Schltr. [71A]

One of the tiniest orchids found in East Africa but easily recognised by its distinctive habit, its bilaterally compressed leaves arranged in a fan on a short stem. They are also distinctively grooved on the upper surface and rarely reach 3 cm long. Inflorescences produced from the leaf-axils, commonly overtopping the leaves, reach 4 cm in length. They bear many tiny white flowers, about 1.5 mm long, along their length but these are hidden almost entirely by the conspicuous dark bracts.

It is a widespread species from Uganda and Kenya southwards to Zimbabwe but plants in our area are referable to subsp. *picea* Cribb distinguished from the typical subspecies by its broader shorter leaves, smaller flowers, shorter straighter spur only about 0.5 mm long, and darker almost black bracts. It is particularly common in the relict forest patches of the Porotos, Umalila and Undali over about 2 000 m and can be found in flower there during March and April.

Bulbophyllum scaberulum (Rolfe) Bolus [71B]

B.scaberulum is a most distinctive orchid when in flower for its flowers are borne on a flattened leaf-like inflorescence axis. It is a creeping plant with yellow pseudobulbs borne at intervals on a stout rhizome. These pseudobulbs are usually about 5 cm tall and two-leaved at the apex. The leaves are leathery and range from 4.5 to 12 cm long and 1 to 1.9 cm broad. The inflorescences are usually longer than the leaves ranging from 5.5 to 24 cm long. The inturning flowers, about 7 mm long, are commonly green heavily marked with maroon.

It is a widespread species in Africa as far south as South Africa. In the Southern Highlands it grows as an epiphyte on tree branches in mist forests up to about 2 000 m from Imagi Mt, near Iringa, west to the Mbeya region.

The closely allied species *B.sandersonii* (Harv.) Bolus is also found commonly in the forests north of the Lake but may be distinguished by its larger flowers with slender petals which are swollen at the apex.

Bulbophyllum stolzii Schltr. [71C]
A small creeping epiphytic orchid with a slender wiry rhizome bearing small ovoid two-leaved pseudobulbs at 3 to 7 cm intervals. Leaves rather leathery, up to about 3.5 cm long and 0.8 cm broad. Inflorescences emerge from the base of the pseudobulb, up to 9 cm long, bearing 4 to 5 mm long, greenish-white, marked with purple outside and with a yellow lip within.

It grows in the higher altitude rain forests above 1 800 m around the top end of Lake Nyasa often in masses in moss and lichens on the lower trunks and branches of forest trees, for example, on Ngosi Mt, on the Umalila escarpment and above Kiwira.

Cyrtorchis crassifolia Schltr. [71D]
The smallest species of the genus *Cyrtorchis* and also the most distinctive, it is a small epiphyte readily recognised by its distichous, fleshy, glaucous leaves, 2 to 4 cm long, borne on a short stem rarely as much as 7 cm long. Inflorescences 2 to 5 cm long, commonly 2- to several-flowered, the white or creamy white star-shaped flowers with a distinct spur at the base of the lip. In bud they are commonly green.

It is probably a fairly widespread orchid between 1 300 and 1 800 m growing on trees in the *Brachystegia* woods. A typical site is at Ipyana, Idunda in Mbozi district where it is common on the branches of trees in the woodland at 1 550 m. We have also found it in the centre of Iringa on the branches of a planted gum-tree (*Eucalyptus*). It also occurs in neighbouring parts of Malawi and Zambia.

Several other species of *Cyrtorchis* are found in the area. *C.arcuata* (Lindl.) Schltr., a large plant with large well-spaced flowers, is widespread at low and medium altitudes from 600 to 1 600 m flowering from January until April; *C. praetermissa* Summerh., a somewhat smaller species with fleshier leaves and more compact inflorescences of smaller flowers, is common in the Chivanjee and Ipinda woodlands at 600 to 1 000 m flowering in mid-December.

Diaphananthe stolzii Schltr. [71E]
Yet another of the many plants named after the early German missionary Adolph Stolz, this pretty epiphytic orchid is common in the forests on the mountains around the northern end of Lake Nyasa. Its hanging stems, up to 30 cm long, bear along their length distichously arranged, leathery, dark green leaves which are twisted at the base and more or less in one plane and up to 7 cm long and 2 cm broad. The inflorescences, emerging through the leaf-sheaths opposite the leaves, each have about six flowers borne on a zig-zag axis. Flowers translucent, pale yellow-green, about 2 cm across with a slender spur, 1 to 1.4 cm long, produced from the base of the lip.

It can usually be found in flower in April in the forest patches around Kiwira, Mwakaleli and Tukuyu and elsewhere in Unyakyusa above about 2 000

m. Its range extends southwards through Malawi to Eastern Zimbabwe.

Several other species of *Diaphananthe* have also been found in the Southern Highlands but less commonly. *D.rutila,* with inflorescences of many tiny flowers, grows at Iviji, Umalila at 1 700 m where it flowers in early April; *D.pulchella* with short stems and yellow flowers almost as large as those of *D.stolzii* can be seen in flower in November and December at Isuto, Umalila and elsewhere; and *D.microphylla,* a dwarf species with very small flowers and incurved spur, can be found at Mwakaleli, Unyakyusa growing on coffee trees at 1 500 m.

Margelliantha leedalii Cribb [72A]

The genus *Margelliantha* contains only five species of small epiphytic orchids and three of these are confined to the mountains of Tanzania. *M.clavata* Cribb is a rarity from the Usambara Mts and *M.globularis* Cribb is confined to the Uluguru Mts. *M.leedalii,* illustrated here, is the most widespread species having been recorded from the Uluguru Mts and from several places in the Southern Highlands as far west as the Poroto Mts.

It is a small plant, up to about 10 cm tall. Its leaves, characteristically twisted at the base to lie one plane, are up to 8 cm long. The hanging inflorescences are 3- to 7-flowered and reach 6.5 cm long. Each white flower is very delicate, somewhat cup-shaped and about 6 to 7 mm across, bearing in its centre a bright green anther cap and with a club-shaped 6 mm long spur at the back.

M.leedalii is a species found in the mountain forests growing on mossy tree trunks and branches from 1 500 to 2 400 m. We have seen it flowering in February on Mt Rungwe in the Kiwira Forest and it has also been recorded from Imagi Mt east of Iringa. Its small size probably means that it has been overlooked elsewhere and it may be far commoner than is, at present, appreciated.

Polystachya goetzeana Kraenzl. [72B]

Polystachya is one of the largest genera of orchids in East Africa. Nearly all are epiphytic plants but a few also grow on rocks where the moisture is high enough.

P.goetzeana is one of the larger flowered species, always found growing as an epiphyte on trees and bushes in the moist forests of the area. It is an erect or pendulous herb, up to 30 cm long. The clustered pseudobulbs are 5 to 15 mm tall and 5 to 7 mm in diameter, each bearing two to four grasslike arching leaves, up to 22 cm long. The flowers are yellow or yellow-green with a white lip which has a yellow ridge at its base. The sepals are 9 to 14 mm long and the slightly longer lip projects from the top of the flower.

This species is confined to the Southern Highlands but is widespread from Kibao near Iringa across to the forests of the Mbeya highlands at altitudes between 1 850 and 2 500 m, flowering in the dry season from September until November.

Plate 72. A, *Margelliantha leedalii;* B, *Polystachya goetzeana;* C, *P.malilaensis;* D, *Stolzia leedalii;* E, *S.nyassana;* F, *Tridactyle tricuspis*

Polystachya malilaensis Schltr. [72C]

Adolph Stolz collected many new species of orchids in our area. *P.malilaensis,* which he collected on Rungwe Mt by the Kiwira River where it grows as an epiphyte on tree branches by the river, was one of these. It is also found growing on rocks in moist areas on the mountains.

This delicate species reaches 15 cm tall. The conical pseudobulbs, each about 2.5 cm long, grow in small clumps bearing two to three leaves, 4 to 12 cm long and 5 to 17 mm broad. The hairy inflorescence bears a few flowers towards the apex. The flowers are 8 to 12 mm long and commonly greenish or yellow-green with a white glabrous lip.

P.malilaensis is found between 1 590 and 2 300 m throughout the mountains around the top end of Lake Nyasa flowering in February and March. It is also found in Malawi where it is often confused with *P.zambesiaca* Rolfe which is readily distinguished by the glandular hairs on its lip.

Stolzia leedalii Cribb [72D]

About six species of *Stolzia* have been recorded from the Southern Highlands, more than half having been described only recently. No doubt their dwarf habit has led to them being overlooked by most botanists.

Three of the species grow in the montane forests of the Mufindi escarpment. Of these, *S.christopheri* Cribb is the most distinctive with large beetle-like purple flowers appearing at the side of the pseudobulbs. *S.compacta* Cribb subsp. *iringana* Cribb is the smallest of the three and has tiny yellow flowers. The third species, *S.leedalii,* grows with the other two but is easily recognised by its pea-like pseudobulbs, up to 1.2 cm in diameter, which bear one erect slender leaf, 2.5 to 7 cm long and 3 to 4 mm broad. Its large brown flower, 7.5 mm long, is borne at the end of a long slender wiry stalk.

These orchids are all apparently confined to the Mufindi escarpment where they grow on the branches and trunks of trees overhanging streams, such as the Kigogo and Luisenga, at between 1 800 and 1 950 m altitude. *S.leedalii* flowers there in October, *S.compacta* subsp. *iringana* in January and February and *S. christopheri* in March.

Stolzia nyassana Schltr. [72E]

This tiny creeping epiphytic orchid was first collected in the forests at the top end of Lake Nyasa by Adolph Stolz, a German missionary who lived in the area some 70 to 80 years ago. Several other plants are named after this prolific collector including other orchids such as *Nervilia adolphii, N.stolzii, Habenaria stolzii* and *Disa stolzii.*

It is a creeping herb, to 10 cm tall, each shoot producing a small fleshy two-leaved pseudobulb up to 1.3 cm high. The leaves are up to 6 cm long and 1.3 cm broad and between them arises the 6 cm long inflorescence which bears

3 to 11 clustered yellow flowers. These are about 5 mm long and all point the same way.

We have found this pretty orchid growing on mossy branches in the mist forests of the Poroto, Undali and Livingstone Mts, at between 1 700 and 2 200 m, flowering in January and February. Elsewhere, it has been recorded only from the Ukaguru Mts north of Morogoro and from neighbouring Malawi.

Tridactyle tricuspis (Bolus) Schltr. [72F]

Some 30 species of the epiphytic orchid genus *Tridactyle* are found in East Africa. Most of these are straggling long-stemmed plants with short to long inflorescences of small green to orange flowers mostly readily recognised by their characteristically three-lobed lip.

One of the commonest species of the Southern Highlands is *T.tricuspis* which, unlike most species, has a short stem rarely up to 20 cm long. Its leaves are relatively thin in texture and reach 13 cm long and 1 cm broad. Several inflorescences, 6 to 15 cm long, are commonly produced on each plant and these have many distichously arranged flowers along their length. The flowers have a musky scent and are commonly butterscotch yellow, buff or orange in colour. The sepals and petals are 6 to 7.5 mm long and a straight 10 to 14 mm long spur is produced from the back of the flower.

T.tricuspis is widespread in the mountain forests of the Southern Highlands, from Imagi Mt in the east across to the Poroto Mts and Umalila in the west, at altitudes up to 2 800 m. In a typical locality such as the Ndumbi Forest on the Kitulo Plateau, it flowers in February and March but, elsewhere, it can be seen in flower as early as January up until May.

T.tricuspis is also found in southern Zaire and southwards to South Africa but, in northern Tanzania and Kenya, the larger flowered but otherwise similar *T.furcistipes* Summerh. replaces this species in similar habitats.

SELECTED BIBLIOGRAPHY

Agnew, A. 1974. *Upland Kenya wild flowers.* Oxford Univ. Press.

Bjørnstad, A. 1976. *The vegetation of the Ruaha National Park, Tanzania.* 1. Annotated check-list of the plant species. Serengeti Research Inst. Publ. 215:61pp.

Brenan, J.P.M. 1949. *Check-lists of the forest trees and shrubs of the British Empire 5.* Tanganyika Territory pt.2. Imperial Forestry Institute, Oxford.

Condry, W.M. 1967. *Birds and wild Africa.* Collins, London.

Dale, I.R. & P.J.Greenway 1961. *Kenya trees and shrubs.* Buchanan's Kenya Estates Ltd, Nairobi.

Eggeling, W.J. & I.R.Dale 1952. *The indigenous trees of the Uganda Protectorate.* Entebbe, Uganda.

Exell, A.W., A.Fernandes & H.Wild 1960 continuing. *Flora Zambesiaca.* London.

Geesink, R. *et al.* 1981. *Thonner's analytical key to the families of flowering plants.* Leiden Univ. Press, The Hague, Netherlands.

Greenway, P. 1937. *A Swahili dictionary of plant names.* Dar es Salaam.

Grey-Wilson, C. 1980. *Impatiens of Africa,* Balkema, Rotterdam.

Gunn, M. & L.E.Codd 1981. *Botanical Exploration of Southern Africa.* Balkema, Rotterdam.

Hedberg, O. 1968. Conservation of vegetation in Africa south of the Sahara. *Acta Phytogeogr. Suec.* 54:1-320.

Heywood, V.H. 1978. *Flowering plants of the world.* Oxford Univ. Press.

Hora, F.B. 1940. *Check-lists of the forest trees and shrubs of the British Empire 5.* Tanganyika Territory pt.1. Imperial Forestry Institute, Oxford.

Hubbard, C.E. & E.Milne-Redhead 1952 continued. *Flora of tropical East Africa.* London.

Hutchinson, J. & J.M.Dalziel 1954-1972. *Flora of west tropical Africa,* ed.2. London.

Kokwaro, J. 1976. *Medicinal plants of East Africa.* Kampala, Uganda.

Lind, E.M. & M.E.S.Morrison 1974. *East African vegetation.* Longmans, London.

Lind, E.M. & A.C.Tallantire 1962. *Some common flowering plants of Uganda.* Oxford Univ. Press.

Moreau, R.E. 1966. *The bird faunas of Africa and its islands.* Academic Press, London.

Moriarty, A. 1975. *Wild flowers of Malawi.* Purnell, Cape Town.

White, F. 1962. *Forest flora of Northern Rhodesia.* Oxford Univ. Press.

Williamson, G. 1976. *Orchids of south-central Africa.* Dent, London.

INDEX TO BOTANICAL NAMES